计算机网络实验教程

——基于eNSP+Wireshark

张 举 耿海军 编著

电子工业出版社
Publishing House of Electronics Industry
北京·BEIJING

内 容 简 介

计算机网络是高等院校计算机及通信相关专业的重要课程之一，但在学习过程中，读者普遍感觉较抽象、难以理解，其主要原因之一是理论和实践不能很好地结合起来，缺少实践方面的训练。

本书从组网的角度设计实验，在内容上以谢希仁教授编著的《计算机网络（第 8 版）》为依托，按照计算机网络的五层体系结构顺序，由下到上安排实验，并与教材中的相关理论相互印证，让读者能够更好地理解计算机网络的原理，并能获得一定的组网能力，为进一步学习打下坚实的基础。

本书共 25 个实验项目，建议教学课时为 36 学时。

本书可作为谢希仁教授编著的《计算机网络（第 8 版）》的配套实验用书，同时也适合作为高校计算机网络课程的实验用书。

未经许可，不得以任何方式复制或抄袭本书之部分或全部内容。
版权所有，侵权必究。

图书在版编目（CIP）数据

计算机网络实验教程：基于 eNSP+Wireshark / 张举，耿海军编著. —北京：电子工业出版社，2021.9
ISBN 978-7-121-20866-9

Ⅰ. ①计… Ⅱ. ①张… ②耿… Ⅲ. ①计算机网络－高等学校－教材 Ⅳ. ①TP393

中国版本图书馆 CIP 数据核字（2021）第 178249 号

责任编辑：牛晓丽　　　　　　特约编辑：田学清
印　　刷：北京天宇星印刷厂
装　　订：北京天宇星印刷厂
出版发行：电子工业出版社
　　　　　北京市海淀区万寿路 173 信箱　　　邮编：100036
开　　本：787×1092　1/16　　印张：11.25　　字数：288 千字
版　　次：2021 年 9 月第 1 版
印　　次：2024 年 7 月第 7 次印刷
定　　价：39.80 元

凡所购买电子工业出版社图书有缺损问题，请向购买书店调换。若书店售缺，请与本社发行部联系，联系及邮购电话：（010）88254888，88258888。

质量投诉请发邮件至 zlts@phei.com.cn，盗版侵权举报请发邮件至 dbqq@phei.com.cn。

本书咨询联系方式：9616328（QQ）。

前　言

计算机网络是高等院校计算机及通信相关专业的重要课程之一，但在学习过程中，读者普遍感觉较抽象、难以理解，其主要原因之一是理论和实践不能很好地结合起来，缺少实践方面的训练。

本书共 25 个实验项目，建议教学课时为 36 学时。其中，第 2 章由郝践编写，第 3 章由耿海军编写，第 5 章由张振涛编写，其余各章由张举编写。

本书具有如下特点。

（1）作为谢希仁教授编著的《计算机网络（第 8 版）》的配套实验教程，在实验设计及内容上与其紧密结合，适合作为高校计算机网络课程的实验用书。

（2）实验中包含了理论基础知识、常用配置命令、实验步骤。

（3）实验内容按照五层体系结构划分，从下到上设计，以期和理论教学过程紧密贴合，但实际上很难完全在知识结构上区别开，故在教学过程中可灵活把控。

（4）实验针对计算机网络中使用的基本协议进行设计，通过组网并进行抓包来查看协议信息和分组结构，进一步理解网络中的协议，使理论不再枯燥，增加读者的学习兴趣。

（5）本书最后设计了两个组网的综合实验项目，可增加读者组建网络的能力。

在本书实验环境中，组网使用华为网络模拟器 eNSP 1.2.00.500 V100R002C00 完成，也可以在配备华为设备的实验室完成；抓包分析使用 Wireshark 软件。两者结合，比较适合用于教学。

本书是对教学团队的教学经验的总结，感谢教学团队成员的支持和帮助，感谢贾新春、武俊生教授给出很多重要的参考意见。另外还要特别感谢郝志恒、牛晓丽两位编辑，在两位编辑的大力鼓励和支持下本书才得以出版。

由于编者水平有限，书中难免有疏漏及不足之处，恳请广大读者和专家提出宝贵意见（发送邮件至 445387453@qq.com）。

编者
2021 年 6 月

目 录

第1章 概述 ... 1
1.1 网络的体系结构 ... 1
1.1.1 概述 ... 1
1.1.2 网络协议 ... 1
1.1.3 五层协议的体系结构 ... 2
1.2 常用网络命令简介 ... 4
1.2.1 ping 命令 ... 4
1.2.2 ipconfig 命令 ... 7
1.2.3 netstat 命令 ... 10
1.2.4 arp 命令 ... 16
1.2.5 tracert 命令 ... 18
1.2.6 route 命令 ... 20
1.3 eNSP 常用功能及使用方法 ... 21
1.3.1 eNSP 的基本界面 ... 21
1.3.2 选择并添加设备 ... 22
1.3.3 连接设备 ... 22
1.3.4 配置设备 ... 22
1.3.5 Wireshark 抓包软件 ... 23
1.4 eNSP 设备基本配置模式 ... 25

第2章 物理层 ... 26
实验一：双绞线制作 ... 26
实验二：交换机初始配置及其 Console 接口配置 ... 27

第3章 数据链路层 ... 31
实验一：集线器组建局域网 ... 31
实验二：以太网二层交换机原理实验 ... 33
实验三：交换机中交换表的自学习功能 ... 37
实验四：交换机 VLAN Access 实验 ... 42
实验五：交换机 VLAN Trunk 实验 ... 47
实验六：生成树配置 ... 55
实验七：链路聚合配置 ... 58

第 4 章 网络层 63

实验一：路由器 IP 地址配置及直连网络 63

实验二：ARP 协议分析 67

实验三：静态路由与默认路由配置 72

实验四：RIP 路由协议配置 75

实验五：内部网关协议 OSPF 实验 87

实验六：外部网关协议 BGP 实验 103

实验七：以太网三层交换机实现 VLAN 间路由配置 120

实验八：路由器单臂路由实现 VLAN 间通信 124

实验九：PPP 协议配置（点对点信道） 127

实验十：访问控制列表（ACL）实验 132

实验十一：网络地址转换（NAT）实验 137

第 5 章 运输层 144

实验一：TCP 连接实验 144

第 6 章 应用层 147

实验一：远程终端协议（TELNET）实验 147

实验二：动态主机配置协议（DHCP）实验 149

第 7 章 校园网综合实验 153

实验一：校园网综合实验 1 153

实验二：校园网综合实验 2 161

附录 A eNSP 常见问题及解决办法 171

第 1 章　概述

1.1　网络的体系结构

1.1.1　概述

计算机网络是由多种计算机和各类终端通过通信线路连接起来的复杂系统。在这个系统中，计算机型号不一、终端类型各异，加之线路类型、连接方式、同步方式、通信方式的不同，因此给网络中各节点间的通信带来许多不便。在不同计算机系统之间，真正以协同方式进行通信的任务是十分复杂的。为了设计这样复杂的计算机网络，早在最初的 ARPANET 设计时设计人员便提出了分层的方法。用分层来实现网络的结构化设计，每层对应设计相应的协议，实现对应的功能。这样的"分层"可将庞大而复杂的问题转化为若干较小的局部问题，而这些较小的局部问题总是比较易于研究和处理的。

将计算机网络的各层及其协议的集合称为网络的体系结构。计算机网络划分不同的层次或在不同层次中实现不同的功能，意味着网络处于不同的体系结构。

采用这种分层的结构可以带来很多好处，具体如下。

（1）各层之间是独立的。某一层并不需要知道它的下一层是如何实现的，而仅仅需要知道该层间的接口（即界面）所提供的服务。由于每一层只实现一种相对独立的功能，因此可将一个难以处理的复杂问题分解为若干个较小的且较容易处理的局部问题。这样，整个问题的复杂程度就下降了。

（2）灵活性好。当任何一层发生变化时（如技术的变化），只要层间接口关系保持不变，则在这层以上或以下各层均不受影响。

（3）结构上可分割开。各层都可以采用最合适的技术来实现。

（4）易于实现和维护。这种结构使得实现和调试一个庞大而又复杂的系统变得易于处理，因为整个系统已被分解为若干个相对独立的子系统。

（5）能促进标准化工作。因为每一层的功能及其所提供的服务都已有了精确的说明。

1.1.2　网络协议

网络协议简称协议，是为计算机网络中的数据交换建立的规则、标准或约定的集合。为了完成各层所规定的功能，每一层都要设计若干协议。协议是水平的，其所涉及的实体是通信双方的对等实体，双方共同遵守协议，在协议的约定下进行通信，完成协议约定的任务。相反，在自己的计算机上进行一个不需要和网络上其他主机进行通信的操作，尽管也有各种规定，但不能称其为网络协议。

网络协议由以下三个要素组成。

（1）语法：即数据与控制信息的结构或格式。

（2）语义：即需要发出何种控制信息、完成何种动作和做出何种响应。

（3）同步：即事件实现顺序的详细说明。

对协议的描述通常有两种形式，一种是用便于阅读和理解的文字来描述，另一种是用程序代码来描述，但不管采用哪种形式，都需要对协议做出精确的解释。

1.1.3 五层协议的体系结构

为了促进计算机网络的发展，国际标准化组织成立了一个委员会，在现有网络的基础上，提出了不基于具体机型、操作系统或公司的网络体系结构，即 OSI 参考模型，其全称为开放系统互连参考模型（Open System Interconnection Reference Model，OSI/RM）。

OSI 参考模型是一个七层的体系结构，它们由低到高分别是物理层（Physical Layer）、数据链路层（Data Link Layer）、网络层（Network Layer）、运输层（Transport Layer）、会话层（Session Layer）、表示层（Presentation Layer）和应用层（Application Layer）。第一层到第三层属于 OSI 参考模型的低三层，负责创建网络通信连接的链路；第四层到第七层为 OSI 参考模型的高四层，负责端到端的数据通信。

OSI 参考模型是网络技术的基础，尽管概念清楚，理论也比较完整，但由于太复杂等原因，并没有得到市场的认可。真正占领市场并得到广泛应用的是 TCP/IP 体系结构，这是一个四层的结构，上面三层依次为应用层、运输层、网络层，其最下面的网络接口层并没有具体的内容。虽然在当时的环境下 TCP/IP 体系结构非常快速地满足了市场的需求，将一些异构的网络都包容了进来，但其缺少了对物理层和数据链路层内容的约定。谢希仁教授所编著的《计算机网络》中按照五层协议的体系结构来阐述计算机网络的体系结构，描述更加清晰、简洁。三种体系结构的层次划分及对应关系如表 1-1 所示。

表 1-1 三种体系结构层次对应表

OSI 的体系结构	TCP/IP 的体系结构	五层协议的体系结构
7 应用层	应用层 （各种应用层协议，如 DNS、HTTP、SMTP 等）	5 应用层
6 表示层		
5 会话层		
4 运输层	运输层（TCP 或 UDP）	4 运输层
3 网络层	网络层 IP	3 网络层
2 数据链路层	网络接口层 （没有具体内容）	2 数据链路层
1 物理层		1 物理层

下面以五层协议的体系结构为例，对计算机网络各层所要实现的功能做简要解释。

1. 物理层

物理层传输数据的单位是比特，其主要关心的是在连接各种计算机的传输媒体上如何传输比特流，为了达到这个目的，物理层提供了建立、维护和拆除物理链路所需要的机械的、电气的、功能的和规程的特性。比如，规定用多大的电压代表"1"或"0"，以及当发送方发出比特"1"时，在接收方如何识别出这是比特"1"而不是比特"0"；规定一些连接电缆材质、引线的数目，以及电缆接头的几何尺寸、锁紧装置等物理性内容。

需要注意的是，具体的连接媒介，如传输介质等并非物理层的内容。

物理层常见设备有中继器、集线器、网线等。

2. 数据链路层

数据链路层主要解决利用物理地址进行通信的问题，传输数据的单位是帧，帧的格式中包括的信息有：地址信息部分、控制信息部分、数据部分、检验信息部分等。为了完成这一任务，数据链路层需要解决以下三个基本问题。

1）封装成帧

发送方的数据链路层需要将上层传递下来的协议数据单元封装成帧，再将其传递给下面的物理层，而物理层则将其作为比特流传输出去。接收方的物理层收到比特流后，将其交给上面的数据链路层，而数据链路层会按照与发送方对等的协议划分为一个个的帧，再将数据部分交给上面的协议。

2）透明传输

透明传输用来解决帧里面的数据部分因含有与帧定界符相同的内容而被误认为是帧定界符的问题，因为这会导致一个帧被非正常结束，从而产生无效帧。数据链路层协议会采用一些方法破坏掉其中与帧定界符相同的帧，这就是透明传输的含义。

3）差错检测

在物理层传输比特流时，由于信号质量的问题，可能会导致比特流在接收方出现误码现象。因此，数据链路层设计了差错检测的功能，以防止出现差错的帧继续在网络中占用资源，同时检测出差错的帧将被接收方丢弃。

数据链路层常见设备有网卡、网桥、交换机等。

3. 网络层

网络层传输数据的单位是 IP 数据报，也被称为 IP 分组、分组或包。

网络层最主要的功能是路由选择功能。在计算机网络中进行通信的两台计算机之间可能要经过许多个节点和链路，也可能还要经过好几台路由器所连接的通信子网。网络层的任务就是要选择路由，使发送方的运输层所传下来的报文能够按照目的 IP 地址找到接收方，并交付给接收方的运输层。

网络层常见设备有路由器、防火墙、多层交换机等。

4. 运输层

网络层找到对方的 IP 接口，运输层则要进一步找到对方的应用进程。运输层通信的两端是指操作系统中的应用进程，因为真正的通信是操作系统中应用进程之间的通信。

运输层提供了两个主要的协议，即传输控制协议（TCP）和用户数据报协议（UDP）。TCP 用来提供面向连接的、可靠的传输服务，数据传输单位是报文段；而 UDP 用来提供无连接的、尽最大努力的传输服务，数据传输单位是用户数据报。两种协议为上面的应用层提供了不同的传输服务。

5. 应用层

应用层是网络体系结构的最高层，是直接为应用进程提供服务的。常见应用层协议有虚拟终端协议（TELNET）、文件传输协议（FTP）、简单邮件传输协议（SMTP）、超文本传输协议（HTTP）和域名服务（DNS）等。许多应用程序调用了应用层协议的服务。当然，应用层为完成功能，也会向下面的运输层请求服务。

1.2 常用网络命令简介

本节以 Windows 系统下的命令为例对常用网络命令进行说明。

1.2.1 ping 命令

1. 功能

ping 命令是最常用的命令，特别是在组网中。ping 命令基于 ICMP 协议，从源站点执行，向目的站点发送 ICMP 回送请求报文，目的站点在收到报文后向源站点返回 ICMP 回送回答报文，源站点把返回的结果信息显示出来。

ping 命令用来测试站点之间是否可达，若可达，则可进一步判断双方的通信质量，包括稳定性等。

需要注意的是，有些主机为了防止通过 ping 探测，利用防火墙设置了禁止 ping 或者在参数中设置禁止 ping，这样就不能通过 ping 确定该主机是否处于开启状态或者其他情况，ping 就失效了。

有关 ICMP 的详细解释请参考教材《计算机网络（第 8 版）》4.4.2 节。

2. 命令格式

Windows 系统用户可通过执行"开始→运行"命令，并在弹出的窗口中键入"cmd"来打开命令行程序。在命令提示符后，按如下格式输入：

```
ping [-t] [-a] [-n count] [-l size] [-f] [-i TTL] [-v TOS][-r count]
[-s count][[-j host-list] |[-k host-list]][-w timeout][-R][-S srcaddr] [-4]
[-6] 目标主机
```

其中，目标主机可以是 IP 地址或者域名。

3. 命令参数

ping 的命令参数及其含义如下：

➢ -t   ping 指定的主机，直到停止。若要查看统计信息并继续操作 ping，则按下 Ctrl+Break 组合键；若要停止 ping，则按下 Ctrl+C 组合键。
- ➢ -a　　　　　　　　将地址解析成主机名。
- ➢ -n count　　　　　要发送的回显请求数。
- ➢ -l size　　　　　　发送缓冲区大小。
- ➢ -f　　　　　　　　在数据包中设置"不分段"标志(仅适用于 IPv4)。
- ➢ -i TTL　　　　　　生存时间。
- ➢ -v TOS　　　　　　服务类型(仅适用于 IPv4。该设置已不推荐使用，且对 IP 标头中的服务字段类型没有任何影响)。
- ➢ -r count　　　　　记录计数跃点的路由(仅适用于 IPv4)。
- ➢ -s count　　　　　计数跃点的时间戳(仅适用于 IPv4)。
- ➢ -j host-list　　　与主机列表一起的松散源路由(仅适用于 IPv4)。
- ➢ -k host-list　　　与主机列表一起的严格源路由(仅适用于 IPv4)。
- ➢ -w timeout　　　　等待每次回复的超时时间(毫秒)。
- ➢ -R　　　　　　　　使用路由标头测试反向路由(仅适用于 IPv6)。
- ➢ -S srcaddr　　　　要使用的源地址。
- ➢ -4　　　　　　　　强制使用 IPv4。
- ➢ -6　　　　　　　　强制使用 IPv6。

4. 常见用法实验

1）ping www.163.com

在 Windows 系统中使用 ping 命令发送 4 个 ICMP 回送请求（每个请求 32 字节），正常情况下会收到 4 个响应。以下为命令运行情况：

```
C:\Users\zj>ping www.163.com
正在 Ping z163ipv6.v.bsgslb.cn [60.222.11.28] 具有 32 字节的数据:
来自 60.222.11.28 的回复: 字节=32 时间=13ms TTL=58
来自 60.222.11.28 的回复: 字节=32 时间=14ms TTL=58
来自 60.222.11.28 的回复: 字节=32 时间=15ms TTL=58
来自 60.222.11.28 的回复: 字节=32 时间=13ms TTL=58
60.222.11.28 的 Ping 统计信息:
    数据包: 已发送 = 4，已接收 = 4，丢失 = 0 (0% 丢失)，
往返行程的估计时间(以毫秒为单位):
    最短 = 13ms，最长 = 15ms，平均 = 13ms
```

可以看到，ping 命令没有带任何参数，并且返回了 4 个响应。由统计信息可以看出，发送了 4 个请求，收到了 4 个响应，丢失率为 0%；往返行程的最短、最长及平均时间，时间越短，说明联通越好。根据这些信息可初步判断本机和目标主机的联通状态。

用户经常通过 ping 127.0.0.1 来检测本地主机是否正确安装和配置了 TCP/IP。

2）ping -n20 www.163.com

通过这个命令可以自己定义发送的回送请求个数，对测试网络速度很有帮助。比如，该命令可以测试发送 20 个数据包的情况，通过查看返回的平均时间为多少，最短时间为多少，最长时间为多少来衡量网络联通状态。

3）ping -t www.163.com

该命令会一直进行下去，直到按下 Ctrl+C 组合键停止。若要查看统计信息并继续操作 ping，则可以按下 Ctrl+Break 组合键。

4）ping -l 5600 -n 2 www.163.com

在默认情况下，Windows 系统中 ping 发送的数据包大小为 32 字节，该命令设置回送请求个数为 2，数据包的大小为 5600 字节，但需要注意该值最大为 65 500 字节。以下为命令运行情况：

```
C:\Users\zj>ping -l 5600 -n 2 www.163.com
正在 Ping z163ipv6.v.bsgslb.cn [60.222.11.25] 具有 5600 字节的数据:
来自 60.222.11.25 的回复: 字节=5600 时间=28ms TTL=58
来自 60.222.11.25 的回复: 字节=5600 时间=23ms TTL=58
60.222.11.25 的 Ping 统计信息:
    数据包: 已发送 = 2，已接收 = 2，丢失 = 0 (0% 丢失)，
往返行程的估计时间(以毫秒为单位):
    最短 = 23ms，最长 = 28ms，平均 = 25ms
```

5）ping -i 3 www.163.com

该命令设置 ICMP 请求报文中的 TTL 值为 3，这个值在每经过一台路由器时会被减 1，当被减小到 1 时，路由器会将该分组丢弃，造成超时。所以，当 TTL 值太小时，可能会出现本来网络是通的，但由于 TTL 值耗尽而发生超时的现象，故对此要合理判断。以下为命令运行情况：

```
C:\Users\zj>ping -i 3 www.163.com
正在 Ping z163ipv6.v.bsgslb.cn [60.222.11.21] 具有 32 字节的数据:
来自 218.26.125.125 的回复: TTL 传输中过期。
来自 218.26.125.125 的回复: TTL 传输中过期。
来自 218.26.125.125 的回复: TTL 传输中过期。
来自 218.26.125.125 的回复: TTL 传输中过期。
60.222.11.21 的 Ping 统计信息:
    数据包: 已发送 = 4，已接收 = 4，丢失 = 0 (0% 丢失)，
```

可见，该请求并未到达目的主机，显然，这并非是网络不通，而是 TTL 值被耗尽了。

6）**ping -n 1 -r 7 www.163.com**

该命令设置发送 1 个请求分组，最多记录 7 个路由节点。其中，路由节点的数量最大设置为 9，若需要查看更多路由节点，则可使用 tracert 命令，tracert 命令后面会介绍。以下为命令运行情况：

```
C:\Users\zj>ping -n 1 -r 7 www.163.com
正在 Ping z163ipv6.v.bsgslb.cn [60.222.11.29] 具有 32 字节的数据:
来自 60.222.11.29 的回复: 字节=32 时间=165ms TTL=58
    路由: 118.81.238.68 ->
          218.26.122.106 ->
          218.26.125.5 ->
          60.222.6.189 ->
          60.222.10.25 ->
          60.222.11.1 ->
          60.222.11.29
60.222.11.29 的 Ping 统计信息:
    数据包: 已发送 = 1，已接收 = 1，丢失 = 0 (0% 丢失)，
往返行程的估计时间(以毫秒为单位):
    最短 = 165ms，最长 = 165ms，平均 = 165ms
```

如果多运行几次该命令，那么可以发现其经过的路由节点是不完全一样的，这是因为每个 IP 分组都是独立路由的。

1.2.2 ipconfig 命令

1. 功能

ipconfig 命令用于显示、更新和释放网络地址设置，包括 IP 地址、子网掩码、网关和 DNS 服务器设置等。

2. 命令格式

ipconfig 的命令格式如下：

```
ipconfig [/allcompartments] [/? | /all |
 /renew [adapter] | /release [adapter] |
 /renew6 [adapter] | /release6 [adapter] |
 /flushdns | /displaydns | /registerdns |
 /showclassid adapter |
 /setclassid adapter [classid] |
 /showclassid6 adapter |
 /setclassid6 adapter [classid]
```

其中，adapter 为连接名称，允许使用通配符*和?。

3. 命令参数

ipconfig 的命令参数及其含义如下：

- /?   显示帮助消息。
- /all   显示完整配置信息。
- /release   释放指定适配器的 IPv4 地址。
- /release6   释放指定适配器的 IPv6 地址。
- /renew   更新指定适配器的 IPv4 地址。
- /renew6   更新指定适配器的 IPv6 地址。
- /flushdns   清除 DNS 解析程序缓存。
- /registerdns   刷新所有 DHCP 租约并重新注册 DNS 名称。
- /displaydns   显示 DNS 解析程序缓存的内容。
- /showclassid   显示适配器的所有允许的 DHCP 类 ID。
- /setclassid   修改 DHCP 类 ID。
- /showclassid6   显示适配器允许的所有 IPv6 DHCP 类 ID。
- /setclassid6   修改 IPv6 DHCP 类 ID。

4. 常见用法实验

1）ipconfig

在默认情况下，执行该命令仅显示绑定到 TCP/IP 适配器的 IP 地址、子网掩码和默认网关。以下为命令运行情况：

```
C:\Users\zj>ipconfig
无线局域网适配器 无线网络连接：

连接特定的 DNS 后缀 . . . . . . . :
本地链接 IPv6 地址. . . . . . . . : fe80::cf5:4314:2bb0:3b29%15
IPv4 地址 . . . . . . . . . . . . : 192.168.1.7
子网掩码 . . . . . . . . . . . . : 255.255.255.0
默认网关. . . . . . . . . . . . . : 192.168.1.1
```

2）ipconfig /all

执行该命令将显示接口网络详细信息。以下为命令运行情况（以无线网卡接口为例）：

```
C:\Users\zj>ipconfig /all
无线局域网适配器 无线网络连接：

   连接特定的 DNS 后缀 . . . . . . . :
   描述. . . . . . . . . . . . . . . : Intel(R) WiFi Link 1000 BGN
   物理地址. . . . . . . . . . . . . : 74-E5-0B-57-6D-84
```

```
   DHCP 已启用 . . . . . . . . . . . : 是
   自动配置已启用. . . . . . . . . : 是
   本地链接 IPv6 地址. . . . . . . : fe80::cf5:4314:2bb0:3b29%15(首选)
   IPv4 地址 . . . . . . . . . . . . . . : 192.168.1.7(首选)
   子网掩码  . . . . . . . . . . . . . . : 255.255.255.0
   获得租约的时间 . . . . . . . . . : 2020 年 1 月 2 日 20:28:23
   租约过期的时间 . . . . . . . . . : 2020 年 1 月 7 日 15:16:46
   默认网关. . . . . . . . . . . . . . . : 192.168.1.1
   DHCP 服务器 . . . . . . . . . . . : 192.168.1.1
   DHCPv6 IAID . . . . . . . . . . . : 376759563
   DHCPv6 客户端 DUID. . . . . . : 00-01-00-01-18-C8-61-C9-F0-DE-F1-E7-7F-2F
   DNS 服务器 . . . . . . . . . . . . : fe80::1%15
                                        192.168.1.1
   TCPIP 上的 NetBIOS . . . . . : 已启用
```

3）**ipconfig /release**

执行该命令将释放所有适配器的 IP 地址。以下为命令运行情况：

```
C:\Users\zj>ipconfig /release
Windows IP 配置
不能在无线网络连接 2 上执行任何操作，它已断开媒体连接。
以太网适配器 本地连接 2:
   媒体状态 . . . . . . . . . . . . . . . : 媒体已断开
   连接特定的 DNS 后缀 . . . . . :
无线局域网适配器 无线网络连接 2:
   媒体状态 . . . . . . . . . . . . . . . : 媒体已断开
   连接特定的 DNS 后缀 . . . . . :
无线局域网适配器 无线网络连接:
   连接特定的 DNS 后缀 . . . . . :
   本地链接 IPv6 地址. . . . . . . : fe80::cf5:4314:2bb0:3b29%15
   默认网关. . . . . . . . . . . . . . . :
```

4）**ipconfig /renew**

执行该命令将更新所有适配器，重新获得 IP 地址。以下为命令运行情况：

```
C:\Users\zj>ipconfig /renew
Windows IP 配置
不能在 本地连接 2 上执行任何操作，它已断开媒体连接。
不能在 无线网络连接 2 上执行任何操作，它已断开媒体连接。
以太网适配器 本地连接 2:
```

```
        媒体状态  . . . . . . . . . . . . . . . : 媒体已断开
        连接特定的 DNS 后缀 . . . . . . . :
无线局域网适配器 无线网络连接 2:

        媒体状态  . . . . . . . . . . . . . . . : 媒体已断开
        连接特定的 DNS 后缀 . . . . . . . :
无线局域网适配器 无线网络连接:

        连接特定的 DNS 后缀 . . . . . . . :
        本地链接 IPv6 地址. . . . . . . . : fe80::cf5:4314:2bb0:3b29%15
        IPv4 地址 . . . . . . . . . . . . . . : 192.168.1.7
        子网掩码  . . . . . . . . . . . . . . : 255.255.255.0
        默认网关. . . . . . . . . . . . . . . : 192.168.1.1
```

6）ipconfig /flushdns

执行该命令将清空本机 DNS 缓存。以下为命令运行情况：

```
C:\Users\zj>ipconfig /flushdns
Windows IP 配置
已成功刷新 DNS 解析缓存。
```

7）ipconfig /allcompartments /all

执行该命令将显示有关所有接口的详细信息。

另外，release 和 renew 这两个参数只能在向 DHCP 租用 IP 地址的计算机上起作用。release 将所有租用的 IP 地址归还 DHCP，而 renew 则重新租用 DHCP 分配的 IP 地址。当然，如果未指定适配器名称，那么会释放或更新所有绑定到 TCP/IP 的适配器的 IP 地址租约。

1.2.3　netstat 命令

1. 功能

netstat 命令是 Windows 系统提供的用于查看与 TCP、IP、UDP 和 ICMP 相关的统计数据的网络工具，并能检验本机各接口的网络连接情况。

2. 命令格式

netstat 的命令格式如下：

```
netstat [-a] [-b] [-e] [-f] [-n] [-o] [-p proto] [-r] [-s] [-t] [interval]
```

3. 命令参数

netstat 的命令参数及其含义如下：

> -a 　　　　显示所有连接和侦听接口。
> -b 　　　　显示在创建每个连接或侦听接口时涉及的可执行程序。在某些情况下，已知可

执行程序承载多个独立的组件,则显示创建连接或侦听接口时涉及的组件序列。可执行程序的名称位于底部"[]"中,它调用的组件位于顶部,直至达到 TCP/IP。注意,此选项可能很耗时,并且在没有足够权限时可能失败。

- -e　　　　显示以太网统计。此选项可以与-s 选项结合使用。
- -f　　　　显示外部地址的完全限定域名(FQDN)。
- -n　　　　以数字形式显示地址和接口号。
- -o　　　　显示拥有的与每个连接关联的进程 ID。
- -p proto　显示 proto 指定的协议的连接;proto 可以是下列任何一个:TCP、UDP、TCPv6 或 UDPv6。
- -r　　　　显示路由表。
- -s　　　　显示每个协议的统计。
- -p　　　　用于指定默认的子网。
- -t　　　　显示当前连接卸载状态。
- interval　重新显示选定的统计,以及各个显示间暂停的间隔秒数。按 Ctrl+C 组合键停止重新显示统计。若省略,则 netstat 将打印一次当前的配置信息。

4. 常见用法实验

1) netstat -a

执行该命令将显示所有连接和监听接口。

以下为命令运行结果的部分内容。其中,协议指连接所用的协议;本地地址指本机地址及接口;外部地址指远端主机地址及接口,接口也用协议代替;状态指协议所处的状态。

```
C:\Users\zj>netstat -a
活动连接

  协议    本地地址               外部地址                  状态
  TCP    192.168.1.7:51664     123.125.52.61:http        TIME_WAIT
  TCP    192.168.1.7:51666     45:https                  CLOSE_WAIT
  TCP    192.168.1.7:51667     45:https                  CLOSE_WAIT
  TCP    192.168.1.7:51670     203.208.43.100:https      ESTABLISHED
  TCP    192.168.1.7:51671     230:http                  ESTABLISHED
  TCP    192.168.1.7:51672     220.181.38.156:http       ESTABLISHED
  TCP    192.168.1.7:51673     230:http                  ESTABLISHED
  TCP    192.168.1.7:51674     tsa01s07-in-f10:https     SYN_SENT
  TCP    192.168.1.7:51675     tsa01s09-in-f14:https     SYN_SENT
  TCP    192.168.1.7:51676     39.96.128.53:http         ESTABLISHED
  TCP    192.168.1.7:51677     119.188.96.39:http        ESTABLISHED
  TCP    192.168.1.7:51678     221.7.140.182:http        LAST_ACK
  UDP    [fe80::cf5:4314:2bb0:3b29%15]:1900    *:*
```

```
UDP    [fe80::cf5:4314:2bb0:3b29%15]:2177     *:*
UDP    [fe80::cf5:4314:2bb0:3b29%15]:50506    *:*
```

如果系统正在运行 P2P 类型的应用，如一些下载类的软件，那么这些应用会不断地与外部地址建立 TCP 连接，从而获取下载资源。在这种情况下，在本命令中就会发现大量本地接口正在与外部地址建立 TCP 连接，请读者自行测试。

关于 TCP、UDP 及接口的内容请参考教材《计算机网络（第 8 版）》第 5 章的内容。

2）netstat -n

执行该命令后将会以数字形式显示地址和接口号，如-a 参数中的主机名在这里会被显示成 IP 地址。

在测试命令前，也可以先访问一些 Web 站点，再运行本命令，观察其中的活动连接。以下为命令运行结果的部分内容：

```
C:\Users\zj>netstat -n
活动连接

协议    本地地址              外部地址                状态
TCP     192.168.1.7:50804     203.119.129.47:443      ESTABLISHED
TCP     192.168.1.7:50805     42.236.37.156:80        ESTABLISHED
TCP     192.168.1.7:50806     42.236.38.71:80         ESTABLISHED
TCP     192.168.1.7:50828     42.236.37.155:80        ESTABLISHED
TCP     192.168.1.7:50835     223.167.166.52:80       ESTABLISHED
TCP     192.168.1.7:51432     60.222.11.25:443        ESTABLISHED
TCP     192.168.1.7:51540     111.206.63.21:80        TIME_WAIT
TCP     192.168.1.7:51541     218.26.34.45:443        CLOSE_WAIT
TCP     192.168.1.7:51542     218.26.34.45:443        CLOSE_WAIT
TCP     192.168.1.7:51552     211.144.24.78:443       TIME_WAIT
TCP     192.168.1.7:51560     221.204.23.3:443        TIME_WAIT
TCP     192.168.1.7:51561     120.52.30.45:443        ESTABLISHED
TCP     192.168.1.7:51562     221.204.13.129:443      ESTABLISHED
TCP     192.168.1.7:51564     211.144.24.235:443      ESTABLISHED
```

3）netstat -e

该命令用于显示关于以太网的统计数据。它列出的项目包括传送的数据包的总字节数、错误数、删除数、数据包的数量和广播的数量。这些统计数据既有发送的数据包数量，也有接收的数据包数量。

这个参数选项可以用来统计一些基本的网络流量。以下为命令运行情况：

```
C:\Users\zj>netstat -e
接口统计
```

	接收的	发送的
字节	268634844	51183852
单播数据包	394434	344040
非单播数据包	5460	18561
丢弃	0	0
错误	0	0
未知协议	0	

4）netstat -s

该命令能够按照各个协议分别显示其统计数据，在默认情况下，显示 IP、IPv6、ICMP、ICMPv6、TCP、TCPv6、UDP 和 UDPv6 的统计。如果应用程序（如 Web 浏览器）运行速度比较慢，或者不能显示 Web 网页之类的数据，那么就可以用该命令来查看所显示的信息，仔细查看统计数据的各行，找到出错的关键字，进而确定问题所在。

以下为命令运行结果的部分内容：

```
C:\Users\zj>netstat -s
IPv4 的统计信息

  接收的数据包                = 64377
  接收的标头错误              = 39
  接收的地址错误              = 0
  转发的数据包                = 0
  接收的未知协议              = 0
  丢弃的接收数据包            = 2460
  传送的接收数据包            = 67616
  输出请求                    = 76260
  路由丢弃                    = 0
  丢弃的输出数据包            = 141
  输出数据包无路由            = 16
  需要重新组合                = 3
  重新组合成功                = 1
  重新组合失败                = 0
  数据报分段成功              = 0
  数据报分段失败              = 0
  分段已创建                  = 0
IPv4 的 TCP 统计信息

  主动开放                    = 2955
  被动开放                    = 14
  失败的连接尝试              = 414
  重置连接                    = 274
```

```
    当前连接                  = 10
    接收的分段                = 76398
    发送的分段                = 56839
    重新传输的分段            = 11459
IPv4 的 UDP 统计信息
    接收的数据包              = 4468
    无接口                    = 2452
    接收错误                  = 0
    发送的数据包              = 7720
IPv6 的 UDP 统计信息
    接收的数据包              = 1274
    无接口                    = 1221
    接收错误                  = 0
    发送的数据包              = 2871
```

5）netstat -r

该命令可以显示关于路由表的信息，除了显示有效路由，还显示当前有效的连接。关于路由的知识请参考教材《计算机网络（第 8 版）》第 4 章的内容。

以下为命令运行结果的部分内容：

```
C:\Users\zj>netstat -r
===========================================================================
接口列表
21...00 ff d0 c2 0e 4d ......Sangfor SSL VPN CS Support System VNIC
17...74 e5 0b 57 6d 85 ......Microsoft Virtual WiFi Miniport Adapter #2
16...74 e5 0b 57 6d 85 ......Microsoft Virtual WiFi Miniport Adapter
15...74 e5 0b 57 6d 84 ......Intel(R) WiFi Link 1000 BGN
22...00 00 00 00 00 00 00 e0 Microsoft ISATAP Adapter #2
24...00 00 00 00 00 00 00 e0 Microsoft ISATAP Adapter #3
===========================================================================

IPv4 路由表
===========================================================================
活动路由：
网络目标          网络掩码          网关            接口          跃点数
0.0.0.0           0.0.0.0           192.168.1.1     192.168.1.7    25
127.0.0.0         255.0.0.0         在链路上        127.0.0.1      306
127.0.0.1         255.255.255.255   在链路上        127.0.0.1      306
127.255.255.255   255.255.255.255   在链路上        127.0.0.1      306
192.168.1.0       255.255.255.0     在链路上        192.168.1.7    281
```

```
    192.168.1.7          255.255.255.255       在链路上           192.168.1.7        281
    192.168.1.255        255.255.255.255       在链路上           192.168.1.7        281
    192.168.182.0        255.255.255.0         在链路上           192.168.182.1      276
    224.0.0.0            240.0.0.0             在链路上           192.168.182.1      276
    255.255.255.255      255.255.255.255       在链路上           127.0.0.1          306
    255.255.255.255      255.255.255.255       在链路上           192.168.1.7        281
    255.255.255.255      255.255.255.255       在链路上           192.168.246.1      276
    255.255.255.255      255.255.255.255       在链路上           192.168.182.1      276
===========================================================================
永久路由:
  网络地址              网络掩码              网关                  跃点数
  0.0.0.0               0.0.0.0               10.50.9.254           默认
===========================================================================

IPv6 路由表
===========================================================================
活动路由:
  跃点数  网络目标            网关
   1      306  ::1/128        在链路上
   15     281  fe80::/64      在链路上
===========================================================================
永久路由:
  无
```

6）netstat -p tcp

该命令可以显示 TCP 连接。-p 后面的参数可以是下列任何一个：TCP、UDP、TCPv6 或 UDPv6。

以下为命令运行结果的部分内容：

```
C:\Users\zj>netstat -p tcp
活动连接

  协议      本地地址              外部地址                状态
  TCP       127.0.0.1:5357        zj-PC:53035             TIME_WAIT
  TCP       127.0.0.1:5357        zj-PC:53039             TIME_WAIT
  TCP       127.0.0.1:52498       zj-PC:54530             ESTABLISHED
  TCP       127.0.0.1:52499       zj-PC:52500             ESTABLISHED
  TCP       127.0.0.1:52500       zj-PC:52499             ESTABLISHED
  TCP       127.0.0.1:54530       zj-PC:52498             ESTABLISHED
  TCP       192.168.1.7:52501     223.167.166.52:http     ESTABLISHED
  TCP       192.168.1.7:52504     hn:http                 ESTABLISHED
```

```
TCP    192.168.1.7:52505      hn:http                    ESTABLISHED
TCP    192.168.1.7:52547      203.119.218.69:https       ESTABLISHED
TCP    192.168.1.7:53024      45:https                   CLOSE_WAIT
TCP    192.168.1.7:53034      203.208.50.167:https       ESTABLISHED
TCP    192.168.1.7:53040      tsa01s07-in-f14:https      SYN_SENT
TCP    192.168.1.7:53041      tsa01s07-in-f14:https      SYN_SENT
```

7）netstat -s -p tcp

该命令可以显示当前 TCP 连接，并对 TCP 协议进行统计。-p 后面的参数可以是下列任何一个：IP、IPv6、ICMP、ICMPv6、TCP、TCPv6、UDP 或 UDPv6。

以下为命令运行结果的部分内容：

```
C:\Users\zj>netstat -s -p tcp
IPv4 的 TCP 统计信息

  主动开放          = 3893
  被动开放          = 221
  失败的连接尝试     = 791
  重置连接          = 327
  当前连接          = 9
  接收的分段        = 105146
  发送的分段        = 77375
  重新传输的分段     = 20766

活动连接

  协议   本地地址              外部地址                   状态
  TCP    192.168.1.7:52547     203.119.218.69:https       ESTABLISHED
  TCP    192.168.1.7:53049     45:https                   CLOSE_WAIT
  TCP    192.168.1.7:53057     106.11.250.27:https        TIME_WAIT
  TCP    192.168.1.7:53062     tsa01s08-in-f46:https      SYN_SENT
  TCP    192.168.1.7:53063     tsa01s08-in-f46:https      SYN_SENT
```

1.2.4　arp 命令

1. 功能

arp 命令用来显示和修改 IP 地址与物理地址之间的映射关系，该映射关系用表来表示即为 IP 地址到物理地址之间的转换表，该转换表保存在本地 ARP 缓存中。

2. 命令格式

arp 的命令格式如下：

```
ARP -s inet_addr eth_addr [if_addr]
ARP -d inet_addr [if_addr]
ARP -a [inet_addr] [-N if_addr] [-v]
```

3. 命令参数

arp 的命令参数及其含义如下：

- ➢ -a   通过询问当前协议数据，显示当前 ARP 项。
- ➢ -g   与 -a 相同。
- ➢ -v   在详细模式下显示当前 ARP 项。所有无效项和环回接口上的项都将显示。
- ➢ inet_addr   指定 IP 地址。
- ➢ -N if_addr   显示 if_addr 指定的网络接口的 ARP 项。
- ➢ -d   删除 inet_addr 指定的主机。inet_addr 可以是通配符*，以删除所有主机。
- ➢ -s   添加主机并且将 IP 地址 inet_addr 与物理地址 eth_addr 相关联。物理地址是用连字符分隔的 6 个十六进制字节。该项是永久的。
- ➢ eth_addr   指定物理地址。
- ➢ if_addr   若存在，则此项指定地址转换表应修改的接口的 IP 地址；若不存在，则使用第一个适用的接口。

4. 常见用法实验

1）arp -a

执行该命令将显示 ARP 缓存中的 IP 地址和物理地址的对应关系。若不止一个网络接口使用 ARP，则显示每个接口的 ARP 表项。以下为含 3 个接口的 arp 命令运行结果：

```
C:\Users\zj>arp -a
接口: 192.168.1.7 --- 0xf
  Internet 地址         物理地址              类型
  192.168.1.1          90-86-9b-87-bb-00     动态
  192.168.1.4          d4-a1-48-44-f7-3b     动态
  192.168.1.255        ff-ff-ff-ff-ff-ff     静态
  224.0.0.22           01-00-5e-00-00-16     静态
  224.0.0.252          01-00-5e-00-00-fc     静态
  239.255.255.250      01-00-5e-7f-ff-fa     静态
  255.255.255.255      ff-ff-ff-ff-ff-ff     静态
接口: 192.168.182.1 --- 0x13
  Internet 地址         物理地址              类型
  192.168.182.255      ff-ff-ff-ff-ff-ff     静态
  224.0.0.22           01-00-5e-00-00-16     静态
  224.0.0.252          01-00-5e-00-00-fc     静态
```

```
        239.255.255.250       01-00-5e-7f-ff-fa          静态
    接口：192.168.246.1 --- 0x14
        Internet 地址         物理地址                    类型
        192.168.246.255       ff-ff-ff-ff-ff-ff          静态
        224.0.0.22            01-00-5e-00-00-16          静态
        224.0.0.252           01-00-5e-00-00-fc          静态
        239.255.255.250       01-00-5e-7f-ff-fa          静态
```

若只想显示某个指定 IP 地址的 ARP 缓存记录，则可用如下命令：

```
    C:\Users\zj>arp -a 192.168.1.1
    接口：192.168.1.7 --- 0xf
        Internet 地址         物理地址                    类型
        192.168.1.1           90-86-9b-87-bb-00          动态
```

若只想显示某个接口的 ARP 缓存记录，则可用如下命令：

```
    C:\Users\zj>arp -a -n 192.168.1.7
    接口：192.168.1.7 --- 0xf
        Internet 地址         物理地址                    类型
        192.168.1.1           90-86-9b-87-bb-00          动态
        192.168.1.4           d4-a1-48-44-f7-3b          动态
        192.168.1.255         ff-ff-ff-ff-ff-ff          静态
        224.0.0.22            01-00-5e-00-00-16          静态
        224.0.0.252           01-00-5e-00-00-fc          静态
        239.255.255.250       01-00-5e-7f-ff-fa          静态
        255.255.255.255       ff-ff-ff-ff-ff-ff          静态
```

2）arp -s 167.56.85.112 00-1a-00-62-c6-08

执行该命令将在 ARP 缓存中添加一条静态 ARP 条目，请读者自行验证。

3）arp -d 167.56.85.112

执行该命令将删除刚刚添加的 ARP 条目，请读者自行验证。

另外，在一些 Windows 系统中，如 Windows 7 系统，当运行 arp 命令来添加静态记录或删除某记录时，有时会被提示"请求的操作需要提升"，这时需要使用管理员身份运行命令行程序。在"开始"处搜索到命令行程序后，右击并选择"以管理员身份运行"即可。

1.2.5 tracert 命令

1. 功能

tracert 命令用于探测源节点到目的节点之间数据报经过的路径。数据报的 TTL 值在每经过一台路由器的转发后减 1，当 TTL=0 时，则向源节点报告 TTL 超时。利用这个特性，可将第一个数据报的 TTL 值置 1，内部封装无法交付的 UDP 用户数据报，这样，途经的第一台路由器将向源节点报告 TTL 超时，第二个数据报将 TTL 赋值为 2，以此类推，直到到达目的站点或 TTL 达到最大值 255，这样就可以得到沿途的路由器 IP 地址。详细内容请参考教材《计算机网络（第 8 版）》4.4.2 节。

2. 命令格式

tracert 的命令格式如下：

```
tracert [-d] [-h maximum_hops] [-j host-list] [-w timeout][-R] [-S srcaddr]
[-4] [-6] target_name
```

3. 命令参数

tracert 的命令参数及其含义如下：

➢	-d	不将 IP 地址解析成主机名。
➢	-h maximum_hops	搜索目标的最大跃点数。
➢	-j host-list	与主机列表一起的松散源路由(仅适用于 IPv4)。
➢	-w timeout	等待每个回复的超时时间(以毫秒为单位)。
➢	-R	跟踪往返行程路径(仅适用于 IPv6)。
➢	-S srcaddr	要使用的源地址(仅适用于 IPv6)。
➢	-4	强制使用 IPv4。
➢	-6	强制使用 IPv6。

4. 常见用法实验

1）tracert www.163.com

tracert 后面可以跟域名或 IP 地址，默认 TTL 值为 30。以下为命令运行情况：

```
C:\Users\zj>tracert www.163.com
通过最多 30 个跃点跟踪
到 z163ipv6.v.bsgslb.cn [60.222.11.27] 的路由:
  1    3 ms     3 ms     3 ms  192.168.1.1 [192.168.1.1]
  2    7 ms    11 ms     8 ms  1.20.185.183.adsl-pool.sx.cn [183.185.20.1]
  3   12 ms    31 ms    34 ms  149.124.26.218.internet.sx.cn [218.26.124.149]
  4   16 ms    16 ms    13 ms  242.5.222.60.adsl-pool.sx.cn [60.222.5.242]
  5   59 ms    67 ms    67 ms  190.6.222.60.adsl-pool.sx.cn [60.222.6.190]
  6   25 ms    19 ms    18 ms  22.10.222.60.adsl-pool.sx.cn [60.222.10.22]
  7   14 ms    14 ms    13 ms  27.11.222.60.adsl-pool.sx.cn [60.222.11.27]
跟踪完成。
```

tracert 命令的结果清晰地显示了去往目的地所经过的路由,"[]"前面是 IP 地址对应的主机名。从命令运行结果可以看到,封装同一 TTL 值的数据报被发送了 3 次。

2) tracert -h 5 60.222.11.27

该命令设置 TTL 值为 5,读者可运行该命令并观察结果。

1.2.6 route 命令

1. 功能

route 命令用来增加、删除或显示本地路由表。

2. 命令格式

route 的命令格式如下:

```
route [-f] [-p] [-4|-6] command [destination][MASK netmask] [gateway] [METRIC metric][IF inte]
```

3. 命令参数

route 的命令参数及其含义如下:

➤	-f	清除所有网关项的路由表。如果与某个命令结合使用,那么在运行该命令前,应清除路由表。
➤	-p	与 ADD 命令结合使用时,将路由设置为在系统引导期间保持不变。在默认情况下,重新启动后不保存路由。忽略所有其他命令,这始终会影响相应的永久路由。
➤	-4	强制使用 IPv4。
➤	-6	强制使用 IPv6。
➤	command	其中之一:
		PRINT 打印路由
		ADD 添加路由
		DELETE 删除路由
		CHANGE 修改现有路由
➤	destination	指定主机。
➤	MASK	指定下一个参数为"网络掩码"值。
➤	netmask	指定此路由项的子网掩码值。如果未指定,其默认设置为 255.255.255.255。
➤	gateway	指定网关。
➤	interface	指定路由的接口号。
➤	METRIC	指定跃点数,如目标的成本。

4. 常见用法实验

1) route print

该命令效果同 netstat -r 完全一致,不再介绍。

2）route add 10.0.0.0 mask 255.0.0.0 192.168.182.1 if 19

该命令将增加一条目的 IP 地址为 10.0.0.0，掩码为 255.0.0.0 的路由条目。命令结束后，读者可使用 route print 命令查看，可以看到该条目已经被添加到本地路由表里了。

3）route delete 10.0.0.0 mask 255.0.0.0

运行该命令后，刚刚添加的路由条目被删除，读者可自行查看。

需要注意的是，若 route 后面添加命令参数，则需要以管理员身份运行命令行处理程序。

当命令为 route print 或 route delete 时，目标或网关可以为通配符（通配符指定为星号*），否则可能会忽略网关参数。若 Dest 包含一个*或?，则会将其视为 Shell 模式，并且只打印匹配目标路由。*匹配任意字符串，而?匹配任意一个字符。

1.3 eNSP 常用功能及使用方法

1.3.1 eNSP 的基本界面

打开 eNSP，其主界面如图 1-1 所示。

图 1-1　eNSP 主界面

图 1-1 所标示区域解释如下。

（1）菜单栏。此栏中有论坛、官网、设置、帮助按钮，在此区域可以进行一些基本的设置，如界面设置、CLI 设置、字体设置、服务器设置、工具设置等。

（2）主工具栏。此栏提供了一些基本设置，如放大/缩小、开启/停止设备、数据抓包等。

（3）工作区。中间的空白处为工作区，在此区域中可以创建网络拓扑、监视模拟过程、查看各种信息和统计数据。

（4）设备类型选择栏。此库包含不同类型的设备，如路由器、交换机、集线器、无线设备、连线、终端设备和网云等。

（5）具体设备选择栏。此库包含不同设备类型中不同型号的设备，它随着设备类型库的选择级联显示。

（6）设备简介栏。此库有当前设备的简单介绍。

1.3.2 选择并添加设备

在设备类型选择栏中选择好网络设备后，具体设备选择栏会级联显示出各种网络设备。以添加路由器为例，简述添加设备的步骤。首先选择路由器，这时具体设备选择栏中会显示各种型号的路由器；然后用鼠标单击想要添加的路由器，按住左键移到工作区后松开，即可将路由器添加到工作区中。当然也可以先双击想要添加的路由器再单击工作区来连续添加设备，以提高效率。

1.3.3 连接设备

选取合适的线型将设备连接起来。可以根据设备间的不同接口选择特定的线型来连接，选择合适的线型后，在设备上单击，会出现接口选择菜单，选择想连接的接口，然后在另一台设备上做同样的操作就可以将两台设备连接起来了。如果只是想快速地建立网络拓扑而不考虑线型选择时，也可以选择自动连线，这时，系统将自动选择接口连接，但并不推荐这种方法。将鼠标移到选择线型并单击时，会对该线型有简单提示。

各种类型的线，依次为自动选线、双绞线、串行线、光纤等。其中，双绞线用于路由器之间的连线。

设备连接完成后，可以看到各线缆两端有不同颜色的圆点，它们表示的含义如表 1-2 所示。

表 1-2 线缆两端圆点的状态及含义

线缆两端圆点的状态	含义
亮绿色	物理连接准备就绪
红色	物理连接不通，没有信号

1.3.4 配置设备

在工作区中右击路由器，选择设置，打开设备配置对话框。

（1）切换到"视图"选项卡，如图 1-2 所示。

"视图"选项卡用于添加接口模块，每选择一个模块，右方会显示出该模块的说明信息。在实物面板视图上可以看到有空槽，首先单击面板上的电源按钮关闭电源，然后用鼠标左键按住该模块，拖到空槽上即可添加模块，最后打开电源按钮。路由器上常用的串口模块有 2E1-F、2SA 等。

（2）切换到"配置"选项卡，如图 1-3 所示。

"配置"选项卡提供了串口号配置，若启动设备时出现串口号冲突的情况，则在此处更改。

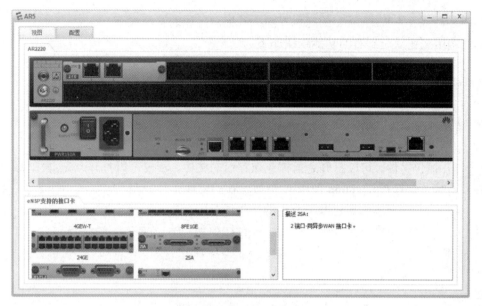

图 1-2 "视图"选项卡

图 1-3 "配置"选项卡

1.3.5 Wireshark 抓包软件

Wireshark（前称 Ethereal）是一个网络数据包分析软件。网络数据包分析软件的功能是截取网络数据包，并尽可能显示出最为详细的网络数据包数据。Wireshark 使用 WinPcap 作为接口，直接与网卡进行数据报文交换。

网络管理员使用 Wireshark 来检测网络问题，网络安全工程师使用 Wireshark 来检查信息安全相关问题，开发者使用 Wireshark 来为新的通信协议除错，普通使用者使用 Wireshark 来学习网络协议的相关知识。

如图 1-4 所示是 Wireshark 的抓包界面，其标示区域的解释如下。

（1）常用按钮从左到右的功能依次如下。

① 列出可用接口。

② 抓包时需要设置的一些选项。一般会保留最后一次的设置结果。

③ 开始新一次的抓包。

④ 暂停抓包。

⑤ 继续进行本次抓包。

⑥ 打开抓包文件。可以打开之前抓包保存后的文件。不仅可以打开 Wireshark 软件保存的文件，而且可以打开 tcpdump 使用-w 参数保存的文件。

⑦ 保存文件。把本次抓包或者分析的结果进行保存。

⑧ 关闭打开的文件。文件被关闭后，就会切换到初始界面。

⑨ 重载抓包文件。

⑩ 打印抓包文件。

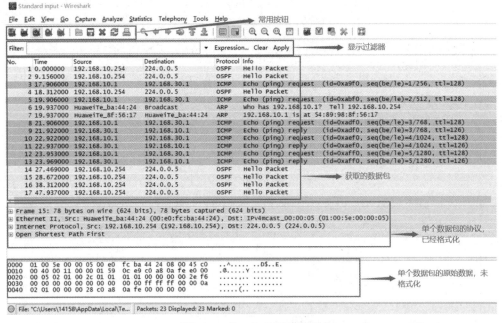

图 1-4　Wireshark 抓包界面

（2）显示过滤器用于捕获文件，用来告诉 Wireshark 只显示那些符合过滤条件的数据包。显示过滤器比捕获过滤器更常用。显示过滤器可以用来过滤不想看到的数据包，但是不会把数据删除。如果想恢复原状，只要把过滤条件删除即可。

（3）获取的数据包，其中颜色有特殊含义，着色规则通过菜单栏中的"视图→着色规则"查看。

（4）数据包的大致结构如下。

第一行：数据包整体概述。

第二行：数据链路层详细信息，主要是双方的物理地址。

第三行：网络层详细信息，主要是双方的 IP 地址。

第四行：不同的报文有不同的内容，图 1-4 中为 OSPF 报文。

1.4 eNSP 设备基本配置模式

在配置设备时，不同模式下可以执行的命令不同。比如，要配置一个接口时，必须进入接口配置模式进行配置，无法在用户模式下对一个接口进行配置。

网络设备的接口也称为接口，本书不做区别。

以路由器为例，每次进入路由器时，首先进入用户模式，在用户模式下不能对路由器进行修改，甚至有些信息都无法查看。当需要对路由器进行其他操作时，只能先进入系统模式，然后再根据需要进入其他模式进行操作。设备配置模式关系如图 1-5 所示。

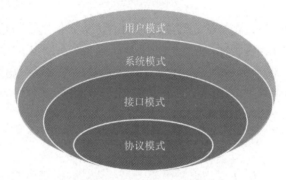

图 1-5　设备配置模式关系

1. 用户模式

交换机启动完成后按下 Enter 键，首先进入的就是用户模式，在此模式下用户将受到极大的限制，只能查看一些统计信息。

2. 系统模式

在用户模式下输入"system-view"（可简写为 sy）命令就可以进入系统模式，用户在此模式下可以查看并修改华为设备的配置，如修改主机名等。

3. 接口模式

在系统模式下输入"interface GigabitEthernet0/0/1"（可简写为 int g0/0/1）命令就可以进入接口模式，用户在此模式下所做的配置都是针对 g0/0/1 这个接口所设定的，如设定 IP 地址等。

4. 协议模式

在系统模式下输入对应协议的名称（如 ospf 1）就可以进入协议模式，用户在此模式下所做的配置都是针对该路由协议所设定的，如 OSPF 宣告网段等。

使用 quit 命令可退回上级模式。不论当前处在哪种模式下，使用 return 命令可直接返回用户模式。

第 2 章 物理层

实验一：双绞线制作

1. 实验目的

掌握制作双绞线的方法。

2. 基本概念

双绞线（Twisted Pair）是结构化布线中最常用的传输媒体之一，它是由两条相互绝缘的铜导线按照一定的规格互相缠绕而成的，根据外部是否有金属屏蔽层分为屏蔽双绞线（STP）和非屏蔽双绞线（UTP）。两条铜导线之所以缠绕在一起是因为这样可以减小信号之间的串扰，如果外界电磁信号在两条铜导线上产生的干扰大小相等而相位相反，那么这个干扰信号就会相互抵消。另外，每对线使用不同颜色以便区分。

由于双绞线价格便宜、安装方便和传输的可靠性高，因此其在短距离数据传输上得到了广泛的应用。

将双绞线和 RJ-45 连接器（水晶头）接在一起，就是我们这里所说的双绞线制作，也即网线制作。双绞线的制作有两种标准，分别是 EIA/TIA-568A 和 EIA/TIA-568B 标准。当双绞线的两端同时是 568A 或 568B 时，为直连双绞线，用来连接不同设备接口；若两边不一样，则为交叉双绞线，用来连接相同设备接口。实际上，现在绝大多数网卡都可以自适应直连和交叉方式进行通信。因此，本实验仅以直连方式为例，两边都按照 T568B 标准制作。

EIA/TIA-568A 标准：绿白、绿、橙白、蓝、蓝白、橙、棕白、棕。

EIA/TIA-568B 标准：橙白、橙、绿白、蓝、蓝白、绿、棕白、棕。

3. 实验步骤

制作双绞线前，要根据拓扑结构设计好双绞线的长度。本实验需要的工具及器件为压线钳（剥线钳）、测线仪、RJ-45 水晶头和 5 类 UTP 双绞线。

（1）用压线钳的剥线刀口将 5 类 UDP 双绞线的外保护套管划开，注意不要将里面的双绞线的绝缘层划破，刀口距 5 类 UDP 双绞线的端头至少 2cm。

（2）将划开的外保护套管剥去，露出 5 类 UTP 双绞线中的 4 对双绞线。

（3）按照 T568B 标准和导线颜色将导线按规定的序号排好，位置参照如图 2-1 所示，将 8 根导线平坦整齐地平行排列，导线间不留空隙。

（4）用压线钳将 8 根导线剪断，注意要剪整齐。剥开的导线长度不可太短，可以先留长一些。

（5）一只手捏住水晶头，将有弹片的一侧向下，有针脚的一端指向远离自己的的方向；另一只手捏平双绞线，最左边是第一脚，最右边是第 8 脚，将剪断的双绞线放入 RJ-45 水晶头中,注意要插到底,并使双绞线的外保护层最后应能够在 RJ-45 水晶头内的凹陷处被压实。

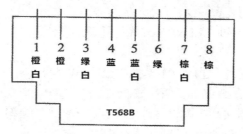

图 2-1 T568B 标准导线位置

(6) 确认正确后,将 RJ-45 水晶头放入压线钳的压头槽内,双手紧握压线钳的手柄,用力压紧,这样,RJ-45 水晶头上的八根针脚便会切破导线绝缘层,和里面的导体压接在一起,就可以传输信号了。

将双绞线的另一端按同样的方法做好。

(7) 测试是否连通。测试时将双绞线两端的 RJ-45 水晶头分别插入主测试仪和远程测试端的 RJ-45 接口,将开关开至"ON"(S 为慢速挡),若主机指示灯从 1 至 8 逐个顺序闪亮,则制作成功,若有灯不亮则说明该灯对应的线不通。测试仪如图 2-2 所示。

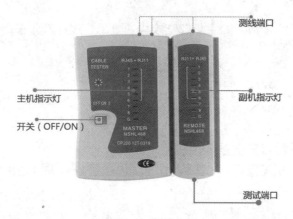

图 2-2 测试仪

实验二:交换机初始配置及其 Console 接口配置

1. 实验目的

(1) 掌握通过 Console 接口对交换机进行配置的方法。
(2) 理解并掌握交换机基本配置。

2. 基本概念

交换机初始配置会进行一些初始的参数配置,如密码、管理 IP 等。

交换机并不配备专门的输入/输出设备,当配置一台新买的交换机时,第一次必须通过 Console 接口来进行。Console 接口是一个串行接口,需要用串行线将其与计算机连接起来,再利用超级终端软件对交换机进行配置，计算机相当于交换机的输入设备。

其他配置方式如下。

（1）TELNET 方式。通过 TELNET 方式远程登录设备进行配置，详见应用层 TELNET 实验。

（2）Web 页面配置。通过一些网管软件或 Web 方式对交换机进行远程配置，优点是使用方便，缺点是有的命令无法在 Web 页面完成。

（3）通过 TFTP 服务器实现对配置文件的保存、下载和恢复等操作，简单方便。

在 eNSP 中，可以直接在命令行进行配置。

3. 实验流程

实验流程如图 2-3 所示。

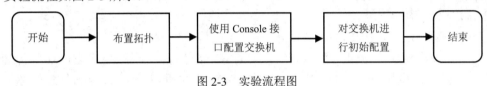

图 2-3　实验流程图

4. 实验步骤

（1）实验拓扑如图 2-4 所示，主机和交换机通过串行线连接起来。

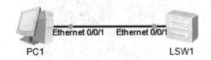

图 2-4　实验拓扑图

（2）使用 Console 接口登录交换机配置。

双击主机，进入主机配置页，在基础配置页面中，可以对主机进行相应的配置，如图 2-5 所示。

图 2-5　主机配置页

由于华为 eNSP 模拟器无法在主机中通过 Console 线配置交换机，因此这里将直接在交换机中配置，但在实际情况中，华为交换机可通过 Console 线连接主机进行配置。

（3）进行交换机基础配置。

系统模式下进行如下基础设置，这里设计交换机管理接口为 VLAN 1，IP 地址为 192.168.1.1/24，交换机命名为 jiaoxue_1，加密用户密码为 Huawei_1，不加密用户密码为 Huawei_2，远程登录密码为 Huawei。下面加黑部分为用户输入：

```
<Huawei>
//打开直接进入用户模式，因为此处没有设置密码，所以直接进入。
<Huawei>system-view
Enter system view, return user view with Ctrl+Z.
//进入系统模式。
[Huawei]sysname jiaoxue_1
//配置交换机名称。
[jiaoxue_1]interface Ethernet 0/0/1
[jiaoxue_1-Ethernet0/0/1]
//进入接口模式。
[jiaoxue_1-Ethernet0/0/1]quit
//从接口模式退回系统模式。
[jiaoxue_1]user-interface console 0
[jiaoxue_1-ui-console0]authentication-mode password
[jiaoxue_1-ui-console0]set authentication password cipher Huawei_1
//配置密码，该密码将被加密。
[jiaoxue_1-ui-console0]set authentication password simple Huawei_2
//配置密码，该密码不加密。
[jiaoxue_1-ui-console0]quit
[jiaoxue_1]aaa
[jiaoxue_1-aaa]local-user admin password cipher Huawei
//虚拟终端密码，用于远程登录。
[jiaoxue_1-aaa]local-user admin service-type telnet
[jiaoxue_1-aaa]quit
[jiaoxue_1]user-interface vty 0 4
[jiaoxue_1-ui-vty0-4]authentication-mode aaa
[jiaoxue_1-ui-vty0-4]quit
[jiaoxue_1]interfaceVlanif 1
//配置管理 IP 地址，二层交换机默认所有接口为 VLAN 1，这里配置 VLAN 1。
[jiaoxue_1-Vlanif1]ip address 192.168.1.1 24
//给 VLAN 1 配置一个 IP 地址，可通过这个地址对交换机进行登录管理。
[jiaoxue_1-Vlanif1]quit
```

```
[jiaoxue_1]quit
<jiaoxue_1>save
The current configuration will be written to the device.
Are you sure to continue?[Y/N]y
//选y，保存退出，直接回车。
Info: Please input the file name ( *.cfg, *.zip ) [vrpcfg.zip]:
Sep  7 2020 23:48:54-08:00 jiaoxue_1 %%01CFM/4/SAVE(l)[1]:The user chose Y when
 deciding whether to save the configuration to the device.
```

第 3 章　数据链路层

实验一：集线器组建局域网

1. 实验目的

（1）理解集线器的工作方式。
（2）理解碰撞域。

2. 集线器的工作方式

最初的以太网是共享总线型的拓扑结构，后来发展为以集线器（Hub）为中心的星形拓扑结构，可以将集线器想象为总线缩短为一点时所成的设备，内部用集成电路代替总线，所以说使用集线器的星形以太网逻辑上仍然是一个总线网。

集线器通常用来直接连接主机，它从一个接口接收信号，并对信号进行整形放大，然后将其从所有其他接口转发出去，是一个有源的设备。集线器工作在物理层，并不识别比特流里面的帧，也不进行碰撞检测，只做简单的物理层的转发，如果信号发生碰撞，那么主机将无法收到正确的比特。

集线器及其所连接的所有主机都属于同一个碰撞域，不同于广播域，碰撞域是指物理层信号的碰撞，是物理层的概念。虽然集线器是一个物理层的设备，但是为便于比较，将此实验放在数据链路层。由于集线器的工作方式非常简单，因此也经常被称为傻 Hub。

详细内容请参考教材《计算机网络（第 8 版）》3.3.3 节。

3. 实验流程

本实验可用一台主机去 ping 另一台主机，并通过 Wireshark 抓包来观察集线器的转发方式，理解碰撞域。实验流程如图 3-1 所示。

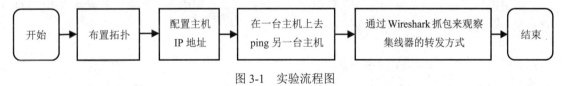

图 3-1　实验流程图

4. 实验步骤

1）单个集线器组网

单个集线器组网的拓扑如图 3-2 所示，各设备的 IP 地址应配置在同一网段，如表 3-1 所示。

表 3-1　各设备的 IP 地址配置

设备名称	IP 地址	子网掩码
PC1	192.168.1.1	255.255.255.0

续表

设备名称	IP 地址	子网掩码
PC2	192.168.1.2	255.255.255.0
PC3	192.168.1.3	255.255.255.0

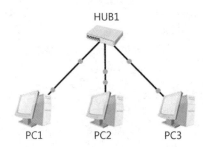

图 3-2 单个集线器组网的拓扑

在命令行下,由 PC1 ping PC3,观察比特流的轨迹。

由图 3-3 和图 3-4 可以看到,集线器将数据包从其他所有接口转发出去,这三台 PC 属于同一碰撞域。

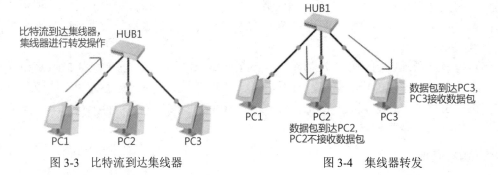

图 3-3 比特流到达集线器　　　　图 3-4 集线器转发

2) 使用集线器扩展以太网

使用集线器扩展以太网的拓扑及数据转发示意如图 3-5 和图 3-6 所示,各设备的 IP 地址应配置在同一网段,具体 IP 地址配置略。

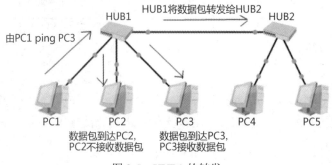

图 3-5 HUB1 的转发

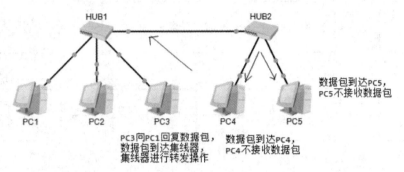

图 3-6　HUB2 的转发

由本实验可以看到，从一台主机所发出的数据包被集线器转发到所有其他主机，即使它们连接在不同的集线器上，这说明所有主机都处在同一个碰撞域中。

实验二：以太网二层交换机原理实验

1．实验目的

（1）理解二层交换机的原理及工作方式。
（2）利用交换机组建小型交换式局域网。

2．交换机的原理及工作方式

交换机是目前局域网中最常用到的组网设备之一，它工作在数据链路层，所以常被称为二层交换机。实际上，交换机有可工作在三层或三层以上层的型号设备，为了表述方便，这里的交换机仅指二层交换机。

数据链路层传输的 PDU（协议数据单元）为帧，不同于工作在物理层的集线器，交换机可以根据帧中的目的 MAC 地址进行有选择的转发，而不是一味地向所有其他接口广播，这依赖于交换机中的交换表。当交换机收到一个帧时，会根据帧里面的目的 MAC 地址去查交换表，并根据结果将其从对应接口转发出去，这种转发方式使得网络的性能得到极大的提升。

交换机的这种转发特性使得接口间可以并行通信，如 1 接口和 2 接口通信时，并不影响 3 接口和 4 接口同时进行通信，当然，前提是交换机必须有足够的背板带宽。

交换机通常有很多接口，如 24 口或 48 口，这些接口在组网中被直接用来连接主机。交换机的接口一般都工作在全双工模式下（不运行 CSMA/CD 协议）。

详细内容请参考教材《计算机网络（第 8 版）》3.4.2 节。

3．实验流程

本实验可用一台主机去 ping 另一台主机，并通过 Wireshark 抓包来观察帧结构，理解交换机的转发过程。实验流程如图 3-7 所示。

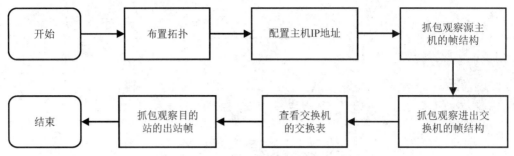

图 3-7 实验流程图

4. 实验步骤

（1）交换机的工作原理。实验拓扑如图 3-8 所示，由 PC1 ping PC2，然后右击 PC1 的 Ethernet0/0/1 接口的绿色标记，单击"开始抓包"。

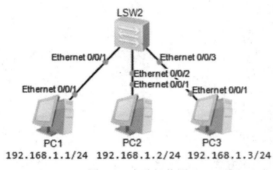

图 3-8 实验拓扑图

在 Wireshark 中观察 PC1 中封装的帧结构，特别是源 MAC 地址和目的 MAC 地址。抓包结果如下：

```
    192.168.1.1    192.168.1.2    ICMP    Echo    (ping)    request    (id=0xf8ec,
seq(be/le)=1/256, ttl=128)    //ping 命令的第一个请求报文。
    Frame 3: 74 bytes on wire (592 bits), 74 bytes captured (592 bits)
    Ethernet   II,   Src:   HuaweiTe_90:41:ef   (54:89:98:90:41:ef),   Dst:
HuaweiTe_f0:0d:e2 (54:89:98:f0:0d:e2)
        Destination: HuaweiTe_f0:0d:e2 (54:89:98:f0:0d:e2) //PC2 的 MAC 地址。
            Address: HuaweiTe_f0:0d:e2 (54:89:98:f0:0d:e2)
            .... ...0 .... .... .... .... = IG bit: Individual address (unicast)
            //第一字节最低位为 0，表明为单个站的地址。
            .... ..0. .... .... .... .... = LG bit: Globally unique address (factory
default)    //第一字节倒数第二位为 0，表明为全球唯一地址，出厂默认。
        Source: HuaweiTe_90:41:ef (54:89:98:90:41:ef)    //PC1 的 MAC 地址。
            Address: HuaweiTe_90:41:ef (54:89:98:90:41:ef)
            .... ...0 .... .... .... .... = IG bit: Individual address (unicast)
            .... ..0. .... .... .... .... = LG bit: Globally unique address (factory
default)
```

```
Type: IP (0x0800)
//类型字段,0x0800 表明里面封装了 IP 数据报。需要说明的是,ping 命令使用了 ICMP 报文,而
//ICMP 报文会被封装为一个 IP 数据报,并被送到目的地,这里所指的就是封装了 ICMP 的 IP 数据报。
Internet Protocol, Src: 192.168.1.1 (192.168.1.1), Dst: 192.168.1.2
(192.168.1.2)
Internet Control Message Protocol

    192.168.1.2 192.168.1.1  ICMP  Echo (ping) reply    (id=0xf8ec, seq(be/le)=1/256,
ttl=128)
//对前面第一个请求报文的回复。
    192.168.1.1 192.168.1.2  ICMP  Echo (ping) request  (id=0xf9ec, seq(be/le)=2/512,
ttl=128)
    192.168.1.2 192.168.1.1  ICMP  Echo (ping) reply    (id=0xf9ec, seq(be/le)=2/512,
ttl=128)
...(略)
```

(2) ping 包的流向如图 3-9 所示。抓取交换机 LSW2 中 Ethernet0/0/1 接口的进站帧和 Ethernet0/0/2 接口的出站帧。

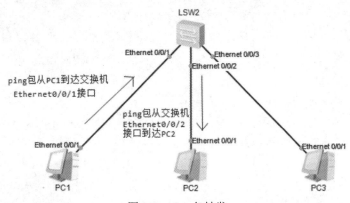

图 3-9 ping 包转发

Ethernet0/0/1 接口的进站帧的抓包结果如下:

```
    192.168.1.1 192.168.1.2 ICMP  Echo (ping) request  (id=0xf8ec, seq(be/le)=1/256,
ttl=128)
    Frame 3: 74 bytes on wire (592 bits), 74 bytes captured (592 bits)
    Ethernet II, Src: HuaweiTe_90:41:ef (54:89:98:90:41:ef), Dst: HuaweiTe_f0:0d:e2
(54:89:98:f0:0d:e2)
        Destination: HuaweiTe_f0:0d:e2 (54:89:98:f0:0d:e2)
            Address: HuaweiTe_f0:0d:e2 (54:89:98:f0:0d:e2)
                .... ...0 .... .... .... .... = IG bit: Individual address (unicast)
                .... ..0. .... .... .... .... = LG bit: Globally unique address (factory
default)
```

```
        Source: HuaweiTe_90:41:ef (54:89:98:90:41:ef)
            Address: HuaweiTe_90:41:ef (54:89:98:90:41:ef)
            .... ...0 .... .... .... .... = IG bit: Individual address (unicast)
            .... ..0. .... .... .... .... = LG bit: Globally unique address (factory
default)
        Type: IP (0x0800)
    Internet Protocol, Src: 192.168.1.1 (192.168.1.1), Dst: 192.168.1.2
(192.168.1.2)
    Internet Control Message Protocol
```

Ethernet0/0/2 接口的出站帧的抓包结果如下：

```
    192.168.1.1 192.168.1.2 ICMP Echo (ping) request (id=0x8bf1, seq(be/le)=1/256,
ttl=128)
    Frame 1: 74 bytes on wire (592 bits), 74 bytes captured (592 bits)
    Ethernet II, Src: HuaweiTe_90:41:ef (54:89:98:90:41:ef), Dst:
HuaweiTe_f0:0d:e2 (54:89:98:f0:0d:e2)
        Destination: HuaweiTe_f0:0d:e2 (54:89:98:f0:0d:e2)
            Address: HuaweiTe_f0:0d:e2 (54:89:98:f0:0d:e2)
            .... ...0 .... .... .... .... = IG bit: Individual address (unicast)
            .... ..0. .... .... .... .... = LG bit: Globally unique address (factory
default)
        Source: HuaweiTe_90:41:ef (54:89:98:90:41:ef)
            Address: HuaweiTe_90:41:ef (54:89:98:90:41:ef)
            .... ...0 .... .... .... .... = IG bit: Individual address (unicast)
            .... ..0. .... .... .... .... = LG bit: Globally unique address (factory
default)
        Type: IP (0x0800)
    Internet Protocol, Src: 192.168.1.1 (192.168.1.1), Dst: 192.168.1.2
(192.168.1.2)
    Internet Control Message Protocol
```

可以发现，不论是进交换机还是出交换机的帧，其源 MAC 地址和目的 MAC 地址都没有被改变，说明尽管每个交换机接口都有各自的 MAC 地址，但进出交换机接口并不会改变帧中的源 MAC 地址和目的 MAC 地址。

该帧被交换机从 Ethernet0/0/2 接口转发到 PC2，之所以没有从 Ethernet0/0/3 接口转发出去，是因为交换机是根据交换表来转发以太网帧的，这也是其和集线器的主要区别。

（3）查看交换机的交换表。进入交换机 CLI 界面，在系统模式中查看交换机的交换表并进行印证，如下所示：

```
    [LSW2]display mac-address                    //显示交换机的交换表。
```

```
MAC address table of slot 0:
---------------------------------------------------------------
MAC Address      VLAN/ PEVLAN CEVLAN Port      Type      LSP/LSR-ID
VSI/SI                                         MAC-Tunnel
---------------------------------------------------------------
5489-9890-41ef  1      -      -      Eth0/0/1  dynamic   0/-
5489-98f0-0de2  1      -      -      Eth0/0/2  dynamic   0/-
//交换表中的记录。
Total matching items on slot 0 displayed = 2
```

（4）观察 PC2 中的进站帧和出站帧，可以看到其出站和进站的 MAC 地址已经相反了，出站帧是 ping 命令对 PC1 的回答，将被发往 PC1。其出站帧如下：

```
Frame 4: 74 bytes on wire (592 bits), 74 bytes captured (592 bits)
    Ethernet    II,   Src:    HuaweiTe_f0:0d:e2   (54:89:98:f0:0d:e2),   Dst:
HuaweiTe_90:41:ef (54:89:98:90:41:ef)
  Destination: HuaweiTe_90:41:ef (54:89:98:90:41:ef)    //PC1 的 MAC 地址。
        Address: HuaweiTe_90:41:ef (54:89:98:90:41:ef)
        .... ...0 .... .... .... .... = IG bit: Individual address (unicast)
        .... ..0. .... .... .... .... = LG bit: Globally unique address (factory
default)
  Source: HuaweiTe_f0:0d:e2 (54:89:98:f0:0d:e2)         //PC2 的 MAC 地址。
        Address: HuaweiTe_f0:0d:e2 (54:89:98:f0:0d:e2)
        .... ...0 .... .... .... .... = IG bit: Individual address (unicast)
        .... ..0. .... .... .... .... = LG bit: Globally unique address (factory
default)
    Type: IP (0x0800)
  Internet   Protocol,   Src:   192.168.1.2   (192.168.1.2),   Dst:   192.168.1.1
(192.168.1.1)
  Internet Control Message Protocol
```

在这种拓扑下，只要主机的 IP 地址在同一网段，主机之间就可以两两 ping 通。这种拓扑通常用来组建一些小型网络，如覆盖一间办公室或宿舍的交换式网络。

实际上，这里 PC1 的 ping 命令要使用 ARP 协议来获得 PC2 的 MAC 地址，并将其封装在自己的 MAC 层的帧中，这里先将其透明处理。

实验三：交换机中交换表的自学习功能

1. 实验目的

理解二层交换机中交换表的自学习功能。

2. 基本概念

交换机可以即插即用，不需要人工配置交换表，交换表的建立是通过交换机自学习得到的。其主要思路为主机 A 封装的帧从交换机的某个接口进入，那么，也可以从该接口到达主机 A。这样，当交换机收到一个帧时，可以将帧中的源 MAC 地址和收到该帧的接口号作为一条记录保存到交换表中，在转发时，若交换表中没有该帧中目的 MAC 地址的记录，则通过广播方式去寻找，即向除该进入接口外的其他所有接口转发。

详细内容请参见教材《计算机网络（第 8 版）》3.4.2 节。

本实验相关命令如下：

```
[LSW1]undo mac-address all          //清空交换机交换表。
```

3. 实验流程

实验流程如图 3-10 所示。

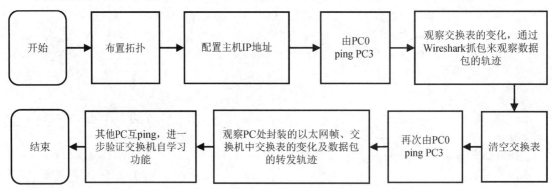

图 3-10 实验流程图

4. 实验步骤

（1）布置拓扑。创建如图 3-11 所示拓扑。

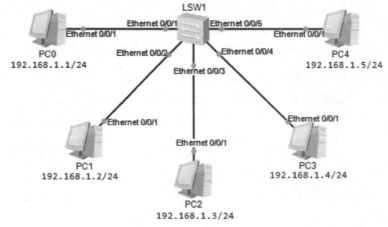

图 3-11 实验拓扑图

（2）执行 ping 命令，观察分组。由 PC0 ping PC3，右击 PC0 的 Ethernet0/0/1 接口，单击"开始抓包"，观察 ARP 协议，该 ARP 分组被封装为以太网广播帧（目的 MAC 地址为全 1），这里暂不考虑 ARP 的工作原理，仅观察 ARP 分组里的源 MAC 地址和目的 MAC 地址，如下所示：

```
No.     Time      Source              Destination    Protocol     Info
 1    0.000000    HuaweiTe_2a:2c:fc    Broadcast      ARP          Who has 192.168.1.3? Tell 192.168.1.1

   Ethernet II, Src: HuaweiTe_2a:2c:fc (54:89:98:2a:2c:fc), Dst: Broadcast (ff:ff:ff:ff:ff:ff)
     Destination: Broadcast (ff:ff:ff:ff:ff:ff)
         Address: Broadcast (ff:ff:ff:ff:ff:ff)
         .... ...1 .... .... .... .... = IG bit: Group address (multicast/broadcast)
         .... ..1. .... .... .... .... = LG bit: Locally administered address (this is NOT the factory default)
     Source: HuaweiTe_2a:2c:fc (54:89:98:2a:2c:fc)    //PC0 的 MAC 地址，将被记入交换表。
         Address: HuaweiTe_2a:2c:fc (54:89:98:2a:2c:fc)
         .... ...0 .... .... .... .... = IG bit: Individual address (unicast)
         .... ..0. .... .... .... .... = LG bit: Globally unique address (factory default)
     Type: ARP (0x0806)
     Trailer: 00000000000000000000000000000000
```

（3）交换机添加交换表记录。如图 3-12 所示，PC0 发送的 ARP 分组到达交换机，此时查看交换机的交换表，如下所示：

```
[LSW1]display mac-address
MAC address table of slot 0:
-------------------------------------------------------------------------
MAC Address    VLAN/   PEVLAN CEVLANPort    Type     LSP/LSR-ID
               VSI/SI                                MAC-Tunnel
-------------------------------------------------------------------------
5489-9898-34ed  1       -      -    Eth0/0/3   dynamic   0/-
5489-982a-2cfc  1       -      -    Eth0/0/1   dynamic   0/-
-------------------------------------------------------------------------
Total matching items on slot 0 displayed = 2
```

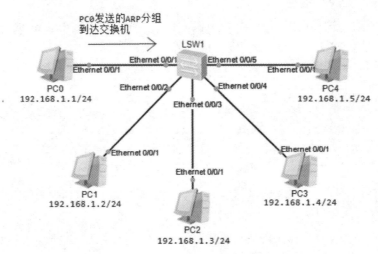

图 3-12　PC0 发送的 ARP 分组到达交换机

实验中利用 ping 命令去访问另一台主机，在 ping 包发出前，网络会先运行 ARP 协议来获得对方主机的 MAC 地址。这样，按照自学习算法，交换机会首先学习到 ARP 分组中的源 MAC 地址（PC0）和对应接口号，并记入交换表。

通过交换表可以看到，PC0 的 MAC 地址已经被交换机自学习到了。需要注意的是，由于模拟器和实验环境的问题，此处交换表显示为两条记录，但这个步骤中实际只有 PC0 的地址被记入交换表。

（4）ARP 分组被交换机广播出去，如图 3-13 所示。但需要注意，此广播属于 ARP 的广播（目的 MAC 地址为全 1），而非交换机由于找不到交换表中的记录所进行的广播。

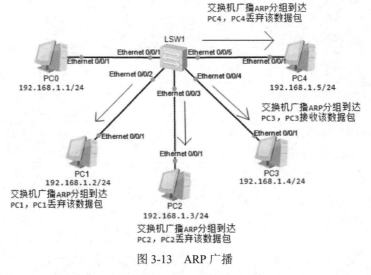

图 3-13　ARP 广播

（5）单击 ARP 的 reply 报文，观察 PC3 的 MAC 地址，如下所示：

```
No.     Time      Source              Destination          Protocol Info
2   0.047000   HuaweiTe_98:34:ed   HuaweiTe_2a:2c:fc      ARP      192.168.1.3 is at 54:89:98:98:34:ed
```

```
Ethernet    II,    Src:    HuaweiTe_98:34:ed    (54:89:98:98:34:ed),    Dst:
HuaweiTe_2a:2c:fc (54:89:98:2a:2c:fc)
    Destination: HuaweiTe_2a:2c:fc (54:89:98:2a:2c:fc)
        Address: HuaweiTe_2a:2c:fc (54:89:98:2a:2c:fc)
        .... ...0 .... .... .... .... = IG bit: Individual address (unicast)
        .... ..0. .... .... .... .... = LG bit: Globally unique address (factory
default)
    Source: HuaweiTe_98:34:ed (54:89:98:98:34:ed)    //PC3 的 MAC 地址。
        Address: HuaweiTe_98:34:ed (54:89:98:98:34:ed)
        .... ...0 .... .... .... .... = IG bit: Individual address (unicast)
        .... ..0. .... .... .... .... = LG bit: Globally unique address (factory
default)
    Type: ARP (0x0806)
    Trailer: 000000000000000000000000000000000000
```

（6）交换机转发 ARP 分组。ARP 分组返回交换机，如图 3-14 所示。此时，按照自学习算法，PC3 的 MAC 地址将被记录入交换表中。

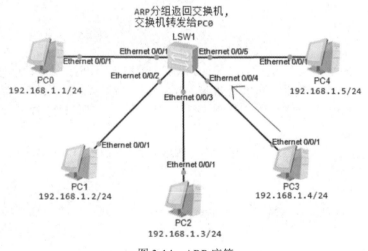

图 3-14　ARP 应答

查看交换机的交换表：

```
[LSW1]display mac-address
MAC address table of slot 0:
-------------------------------------------------------------------------
MAC Address     VLAN/        PEVLAN CEVLAN Port        Type     LSP/LSR-ID
                VSI/SI                   MAC-Tunnel
-------------------------------------------------------------------------
```

```
5489-9898-34ed 1          -      -      Eth0/0/4   dynamic   0/-
5489-982a-2cfc 1          -      -      Eth0/0/1   dynamic   0/-
------------------------------------------------------------------
Total matching items on slot 0 displayed = 2
```

（7）观察交换机的转发。从图 3-15 中可以看到，交换机直接将 ARP 分组由 Ethernet0/0/1 接口转发出去，而不是向其他接口广播，这正是依据交换表转发的结果。

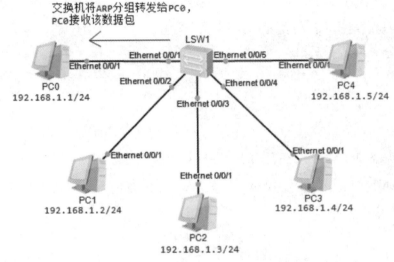

图 3-15　交换机查表后转发

（8）清空交换机的 MAC 地址表，再次由 PC0 ping PC3。此时由于 PC0 的 ARP 缓存中保存有 PC3 的 MAC 地址，因此，PC0 处封装的目的 MAC 地址为 PC3 的 MAC 地址，当帧到达交换机时，因为交换机的交换表中没有该目的地址的记录，所以按照自学习算法将向所有其他接口转发。

ping 命令结束后，再次查看交换机中的交换表，此时交换表中的记录是否发生变化？请读者自行查看。

实验四：交换机 VLAN Access 实验

1. 实验目的

（1）理解二层交换机的缺陷。
（2）理解交换机的 VLAN，掌握其应用场合。
（3）掌握二层交换机 VLAN 的基础配置。

2. VLAN 基础知识

一个二层交换网络属于一个广播域，广播域也可以理解为一个广播帧所能达到的范围。在网络中存在大量的广播，许多协议及应用都是通过广播来完成某种功能的，如 MAC 地址的查询、ARP 协议等。但是，过多的广播包在网络中会发生碰撞，一些广播包会被重传，这样，越来越多的广播包最终会将网络资源耗尽，使得网络性能下降，甚至发生网络瘫痪。

虚拟局域网（Virtual Local Area Network，VLAN）技术可以将一个较大的二层交换网络划分为若干个较小的逻辑网络，每个逻辑网络都是一个广播域，且与具体物理位置无关，因此 VLAN 技术在局域网中被普遍使用。具体来说，VLAN 有如下优点。

（1）控制广播域。每个 VLAN 都属于一个广播域，通过划分不同的 VLAN，广播会被限制在一个 VLAN 内部，这将有效控制广播范围，减小广播对网络的不利影响。

（2）增强网络的安全性。对于有敏感数据的用户组，其可以与其他用户通过 VLAN 隔离，从而减小数据因被广播监听而泄密的可能性。

（3）组网灵活，便于管理。可以按职能部门、项目组或其他管理逻辑来划分 VLAN，便于部门内部的资源共享。由于 VLAN 只是逻辑上的分组网络，因此可以将不同地理位置上的用户划分到同一 VLAN 中。例如，将一幢大楼二层的部分用户和三层的部分用户划到同一 VLAN 中，尽管这些用户的实际网络可能连接在不同的交换机上，地理位置也不同，但却是在一个逻辑网络中，按统一的策略去管理。

交换机中的每个 VLAN 都被赋予一个 VLAN 号，以区别于其他 VLAN，也可以对每个 VLAN 起个有意义的名称，方便理解。

VLAN 划分的方式如下。

（1）基于接口的划分。若将交换机接口划分到某个 VLAN，则连接到该接口上的用户即属于该 VLAN。这种划分方式的优点是简单、方便，缺点是当该用户离开接口时，需要根据情况重新定义新接口的 VLAN。

（2）基于 MAC 地址、网络层协议类型等划分 VLAN。

基于接口的划分方式应用最多，所有支持 VLAN 的交换机都支持这种方式，这里只介绍基于接口的划分。

更多详细内容请参考教材《计算机网络（第 8 版）》3.4.3 节。

本实验配置命令如表 3-2 所示。

表 3-2 本实验配置命令

命令格式	命令说明
vlan vlan-id	创建 VLAN，例如：vlan 10
name vlan-name	给 VLAN 命名（注：模拟器二层/三层交换机暂不支持）
port link-type access	将该接口定义为 access 模式，应用于接口模式下
port link-type access vlan-id	将接口划分到特定 VLAN，应用于接口模式下
display vlan	显示 VLAN 及接口信息
display vlan vlan-id	显示特定 VLAN 信息

3. 实验流程

本实验可用一台主机去 ping 另一台主机，并在不同情况下抓包观察帧的轨迹，理解广播域。实验流程如图 3-16 所示。

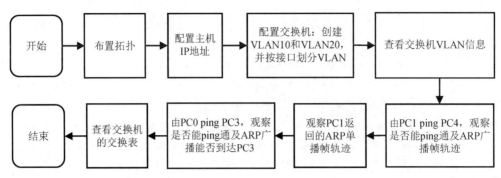

图 3-16 实验流程图

4. 实验步骤

（1）布置拓扑。将主机 IP 地址均设置在 192.168.1.0/24 网段，在交换机中创建 VLAN 10 和 VLAN 20，将 Ethernet0/0/1、Ethernet0/0/2 和 Ethernet0/0/3 接口划入 VLAN 10，Ethernet0/0/4、Ethernet0/0/5 和 Ethernet0/0/6 接口划入 VLAN 20，如图 3-17 所示。PC0、PC1 和 PC4 属于 VLAN 10 的广播域，而 PC2、PC3 和 PC5 属于 VLAN 20 的广播域，观察 VLAN 的作用。

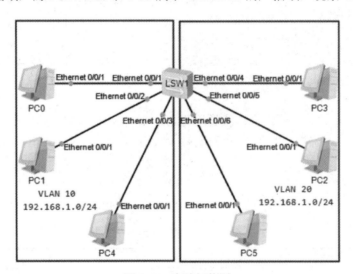

图 3-17 实验拓扑图

（2）配置交换机。对交换机按要求做如下配置：

```
<LSW1>sy
Enter system view, return user view with Ctrl+Z.
[LSW1]vlan 10
[LSW1-VLAN 10]q
[LSW1]vlan 20
[LSW1-VLAN 20]q
[LSW1]int eth 0/0/1
[LSW1-Ethernet0/0/1]port link-type access
[LSW1-Ethernet0/0/1]port default vlan 10
```

```
[LSW1]int eth 0/0/2
[LSW1-Ethernet0/0/2]port link-type access
[LSW1-Ethernet0/0/2]port default vlan 10
[LSW1]int eth 0/0/3
[LSW1-Ethernet0/0/3]port link-type access
[LSW1-Ethernet0/0/3]port default vlan 10
[LSW1]int eth 0/0/4
[LSW1-Ethernet0/0/4]port link-type access
[LSW1-Ethernet0/0/4]port default vlan 20
[LSW1]int eth 0/0/5
[LSW1-Ethernet0/0/5]port link-type access
[LSW1-Ethernet0/0/5]port default vlan 20
[LSW1]int eth 0/0/6
[LSW1-Ethernet0/0/6]port link-type access
[LSW1-Ethernet0/0/6]port default vlan 20
```

经过以上设置后,查看交换机 VLAN 信息:

```
[LSW1]display vlan
The total number of vlans is : 3
--------------------------------------------------------------------
U: Up;         D: Down;       TG: Tagged;      UT: Untagged;
MP: Vlan-mapping;             ST: Vlan-stacking;
#: ProtocolTransparent-vlan;  *: Management-vlan;
--------------------------------------------------------------------
VID  Type    Ports
--------------------------------------------------------------------
1    common  UT:Eth0/0/7(D)   Eth0/0/8(D)    Eth0/0/9(D)   Eth0/0/10(D)
             Eth0/0/11(D)     Eth0/0/12(D)   Eth0/0/13(D)  Eth0/0/14(D)
             Eth0/0/15(D)     Eth0/0/16(D)   Eth0/0/17(D)  Eth0/0/18(D)
             Eth0/0/19(D)     Eth0/0/20(D)   Eth0/0/21(D)  Eth0/0/22(D)
             GE0/0/1(D)       GE0/0/2(D)

10   common  UT:Eth0/0/1(U)   Eth0/0/2(U)    Eth0/0/3(U)

20   common  UT:Eth0/0/4(U)   Eth0/0/5(U)    Eth0/0/6(U)


VID  Status  Property    MAC-LRN Statistics Description
--------------------------------------------------------------------
```

```
1    enable  default         enable  disable   VLAN 0001
10   enable  default         enable  disable   VLAN 0010
20   enable  default         enable  disable   VLAN 0020
```

可见，交换机知道哪些接口属于哪个 VLAN，默认情况下所有接口属于 VLAN 1。

（3）同一 VLAN 广播帧。由 PC0 ping PC4，右击 PC0 的 Ethernet0/0/1 接口，单击"开始抓包"，此处只关注封装 ARP 分组的 MAC 帧。其中第一个是广播帧，这里暂时只关注其广播的属性。由于该分组从 Ethernet0/0/1 接口进入，属于 VLAN 10，因此它将在 VLAN 10 中广播。观察 VLAN 10 的广播域，显然，只有 PC1、PC4 可以收到这个帧，其中 PC1 丢弃该帧，而不属于 VLAN 10 的主机将收不到该广播帧。如下所示，为 PC0 处封装的 ARP 广播帧，注意观察其目的 MAC 地址为广播地址（全 1）：

```
No.    Time       Source              Destination        Protocol   Info
 1   0.000000   HuaweiTe_da:56:9d     Broadcast            ARP      Who has
192.168.1.5? Tell 192.168.1.1

Ethernet II, Src: HuaweiTe_da:56:9d (54:89:98:da:56:9d), Dst: Broadcast
(ff:ff:ff:ff:ff:ff)
  Destination: Broadcast (ff:ff:ff:ff:ff:ff)
      Address: Broadcast (ff:ff:ff:ff:ff:ff)
      .... ...1 .... .... .... .... = IG bit: Group address (multicast/broadcast)
      .... ..1. .... .... .... .... = LG bit: Locally administered address (this
is NOT the factory default)
  Source: HuaweiTe_da:56:9d (54:89:98:da:56:9d)
      Address: HuaweiTe_da:56:9d (54:89:98:da:56:9d)
      .... ...0 .... .... .... .... = IG bit: Individual address (unicast)
      .... ..0. .... .... .... .... = LG bit: Globally unique address (factory
default)
      Type: ARP (0x0806)
      Trailer: 000000000000000000000000000000000000
```

（4）同一 VLAN 单播帧。ARP 广播帧到达 PC4 后，PC4 会向 PC0 回复一个单播帧，根据交换机的交换表自学习算法，PC0 的 MAC 地址会被交换机学习到，所以单播帧将被直接转发到 PC0，而不会向其他接口转发。当然，若转发表中没有该地址，则会在 VLAN 10 中广播该帧。

需要注意的是，前面自学习算法没有提到 VLAN，而交换表是基于 VLAN 的，这是因为交换表的建立过程要依赖广播，而广播只能在同一个 VLAN 内部。当交换机由 Ethernet0/0/3 接口（该接口属于 VLAN 10）收到该 ARP 回复帧后，接下来只会查询 VLAN 10 的交换表，而不会查询 VLAN 20 的交换表。

实际上，属于哪个 VLAN 是交换机的事情，主机端对此毫不知情。主机端封装的帧在进入交换机接口时才被打入 VLAN 标识，而在离开接口时会删掉 VLAN 标识，再交给主机。

（5）不同 VLAN 单播帧。由 PC0 ping PC3，此时 PC0 与 PC3 属于不同 VLAN，交换机从 Ethernet0/0/1 接口收到 ARP 广播帧后，会在 VLAN 10 中广播，PC1 和 PC4 可以收到广播帧，但都被丢弃，而 PC3 则收不到该广播帧，如图 3-18 所示。

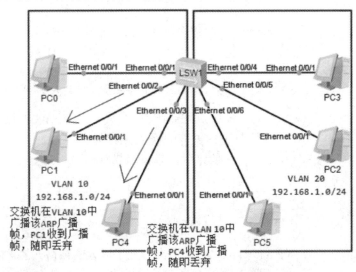

图 3-18　不同 VLAN 单播帧

查看交换机的交换表，注意交换表中 MAC 地址都有 VLAN 标识，目前交换表中没有 VLAN 20 的记录：

```
[LSW1]display mac-address
MAC address table of slot 0:
-----------------------------------------------------------------
MAC Address    VLAN   PEVLAN  CEVLAN Port     Type     LSP/LSR-ID
VSI/SI                                         MAC-Tunnel
-----------------------------------------------------------------
5489-98cc-5803 10     -       -      Eth0/0/3 dynamic  0/-
5489-98da-569d 10     -       -      Eth0/0/1 dynamic  0/-
-----------------------------------------------------------------
Total matching items on slot 0 displayed = 2
```

实验五：交换机 VLAN Trunk 实验

1. 实验目的

（1）理解 VLAN 接口类型 Trunk 的含意。
（2）掌握以太网交换机的 VLAN Trunk 配置。

2. VLAN Trunk 基础知识

首先看一个例子，拥有 VLAN 10 和 VLAN 20 的交换机想要到达另一台拥有相同 VLAN 的交换机时，需要它们在物理上连接两条链路，分别用来承载 VLAN 10 和 VLAN 20 的流量，如图 3-19 所示。

图 3-19 两条链路传播 VLAN 10 和 VLAN 20 数据

Trunk 链路是一条支持多个 VLAN 的点到点链路，允许多个 VLAN 通过该链路到达另一端，如可用一条 Trunk 链路来代替图 3-19 中的两条链路，如图 3-20 所示。显然，对于交换机来说，这种技术节约了接口数量。一般来说，Trunk 链路被设置在交换机之间的连接上。

图 3-20 Trunk 链路传播不同 VLAN 数据

1988 年 IEEE 批准了 802.3ac 标准，这个标准定义了以太网帧格式的扩展。虚拟局域网的帧称为 802.1Q 帧，是在以太网帧格式中插入一个 4 字节的标识符（VLAN 标记），用以指明该帧属于哪一个 VLAN，详见教材《计算机网络（第 8 版）》3.4.3 节。

本实验配置命令如表 3-3 所示。

表 3-3 本实验配置命令

命令格式	命令说明
port link-type trunk	将该接口设置为 Trunk 模式，不理会对方接口是否为 Trunk 模式
port trunk allow-pass vlan vlan-id	将该 vlan-id 添加到 Trunk 中，允许其通过
undo port trunk allow-pass vlan vlan-id	将该 vlan-id 从 Trunk 中移除，不允许其通过
port trunk allow-pass vlan all	Trunk 中允许所有 VLAN 通过
SysnameHost name	设置交换机名称

3. 实验流程

本实验可用一台主机去 ping 另一台主机，并通过 Wireshark 抓包来观察 ICMP 分组的轨迹，理解碰撞域。实验流程如图 3-21 所示。

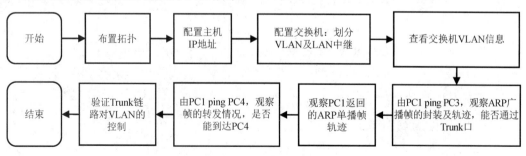

图 3-21 实验流程图

4. 实验步骤

（1）布置拓扑。如图 3-22 所示，拓扑中包含三台交换机 LSW1、LSW2 和 LSW3 及六台主机，将主机 IP 地址均设置在 192.168.1.0/24 网段，交换机 LSW2、LSW3 中创建 VLAN 10 和 VLAN 20。将 LSW2 中的 Ethernet0/0/1 和 Ethernet0/0/2 接口划入 VLAN 10，Ethernet0/0/3 接口划入 VLAN 20。将 LSW3 中的 Ethernet0/0/1 接口划入 VLAN 10，Ethernet0/0/2 和 Ethernet0/0/3 接口划入 VLAN 20。

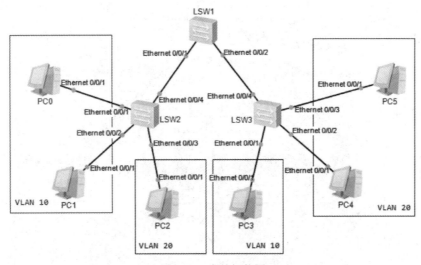

图 3-22 实验拓扑图

（2）配置交换机 VLAN 及接口。创建 VLAN 10 和 VLAN 20，命令如下：

```
<LSW1>sy
Enter system view, return user view with Ctrl+Z.
[LSW1]vlan batch 10 20
Info: This operation may take a few seconds. Please wait for a moment...done.
```

将 LSW2 中的 Ethernet0/0/4 接口配置为 Trunk 模式，允许 VLAN 10 与 VLAN 20 通过，拒绝 VLAN 1 通过，命令如下：

```
[LSW2]int Ethernet0/0/4
[LSW2-Ethernet0/0/4]port link-type trunk
[LSW2-Ethernet0/0/4]port trunk allow-pass vlan 10 20
[LSW2-Ethernet0/0/4]undo port trunk allow-pass vlan 1
```

将 LSW3 中的 Ethernet0/0/4 接口配置为 Trunk 模式，允许 VLAN 10 与 VLAN 20 通过，拒绝 VLAN 1 通过，命令行略。将交换机 LSW1 中的 Ethernet0/0/1 和 Ethernet0/0/2 接口配置为 Trunk 模式，允许 VLAN 10 与 VLAN 20 通过，拒绝 VLAN 1 通过，命令行略。

配置完成后，请查看交换机的 VLAN 信息。

（3）VLAN 10 的广播帧。由 PC1 ping PC3，在 PC1 处生成 ARP 广播分组，该分组被封

装为以太网帧，观察其在不同设备上生成的出站帧及进站帧。需要注意的是，虽然 PC1 被划入 VLAN 10，但在 PC1 处生成的只是一个普通的以太网帧，802.1Q 帧并不是在这里被封装的。

从图 3-23 中可以看到，ARP 广播帧首先到达 LSW2，并由 LSW2 进一步广播到 PC0 和 LSW1，其中 PC0 处的帧被丢弃，广播到 LSW1 处的帧是 802.1Q 帧，即带 VLAN 标记的帧，该帧在交换机 LSW2 转发前被封装，LSW2 的进站帧和出站帧如下。

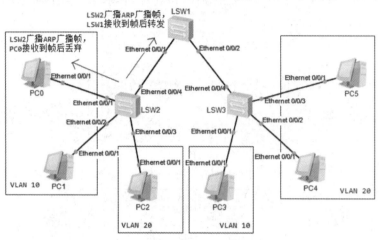

图 3-23　VLAN 10 广播帧

LSW2 的进站帧：

```
    No.     Time          Source              Destination        Protocol       Info
     1  0.000000      HuaweiTe_14:5d:29       Broadcast           ARP         Who has 192.168.1.4?
Tell 192.168.1.2

    Frame 1: 60 bytes on wire (480 bits), 60 bytes captured (480 bits)
    Ethernet II, Src: HuaweiTe_14:5d:29 (54:89:98:14:5d:29), Dst: Broadcast
(ff:ff:ff:ff:ff:ff)
        Destination: Broadcast (ff:ff:ff:ff:ff:ff)
            Address: Broadcast (ff:ff:ff:ff:ff:ff)
            .... ...1 .... .... .... .... = IG bit: Group address (multicast/broadcast)
            .... ..1. .... .... .... .... = LG bit: Locally administered address (this
is NOT the factory default)
        Source: HuaweiTe_14:5d:29 (54:89:98:14:5d:29)
            Address: HuaweiTe_14:5d:29 (54:89:98:14:5d:29)
            .... ...0 .... .... .... .... = IG bit: Individual address (unicast)
            .... ..0. .... .... .... .... = LG bit: Globally unique address (factory
default)
        Type: ARP (0x0806)
        Trailer: 00000000000000000000000000000000
```

```
    Address Resolution Protocol (request)
```

LSW2 的出站帧：

```
No.     Time          Source              Destination     Protocol          Info
 1   0.000000     HuaweiTe_14:5d:29      Broadcast          ARP        Who has 192.168.1.4?
Tell 192.168.1.2

    Frame 1: 64 bytes on wire (512 bits), 64 bytes captured (512 bits)
    Ethernet II, Src: HuaweiTe_14:5d:29 (54:89:98:14:5d:29), Dst: Broadcast
(ff:ff:ff:ff:ff:ff)
      Destination: Broadcast (ff:ff:ff:ff:ff:ff)
          Address: Broadcast (ff:ff:ff:ff:ff:ff)
          .... ...1 .... .... .... .... = IG bit: Group address (multicast/broadcast)
          .... ..1. .... .... .... .... = LG bit: Locally administered address (this
is NOT the factory default)
      Source: HuaweiTe_14:5d:29 (54:89:98:14:5d:29)
          Address: HuaweiTe_14:5d:29 (54:89:98:14:5d:29)
          .... ...0 .... .... .... .... = IG bit: Individual address (unicast)
          .... ..0. .... .... .... .... = LG bit: Globally unique address (factory
default)
      Type: 802.1Q Virtual LAN (0x8100)
    802.1Q Virtual LAN, PRI: 0, CFI: 0, ID: 10     //被打入VLAN标记。
        000. .... .... .... = Priority: Best Effort (default) (0)
        ...0 .... .... .... = CFI: Canonical (0)
        .... 0000 0000 1010 = ID: 10
    Type: ARP (0x0806)
    Trailer: 000000000000000000000000000000000000
    Address Resolution Protocol (request)
```

该帧接着从 LSW1 被广播到 LSW3（LSW3 的进站帧是 802.1Q 帧，出站帧是普通以太网帧），最后被转发到 PC3，请读者自行查看。在这个过程中，交换机的广播都是按照 VLAN 10 的广播域来进行的。这里，PC0、PC1、PC3、LSW1、LSW2、LSW3 都属于 VLAN 10 的广播域。

（4）VLAN 10 的单播帧。这里根据 PC3 返回的 ARP 单播帧来观察、分析单播帧被转发的情况。PC3 生成指向 PC1 的 MAC 地址的以太网单播帧，如下所示：

```
No.     Time          Source              Destination           Protocol      Info
 2   0.046000     HuaweiTe_ff:1f:a6      HuaweiTe_14:5d:29         ARP      192.168.1.4
is at 54:89:98:ff:1f:a6
```

```
    Frame 2: 64 bytes on wire (512 bits), 64 bytes captured (512 bits)
    Ethernet    II,    Src:    HuaweiTe_ff:1f:a6    (54:89:98:ff:1f:a6),    Dst:
HuaweiTe_14:5d:29 (54:89:98:14:5d:29)
        Destination: HuaweiTe_14:5d:29 (54:89:98:14:5d:29)   //PC1 的 MAC 地址。
            Address: HuaweiTe_14:5d:29 (54:89:98:14:5d:29)
                .... ...0 .... .... .... .... = IG bit: Individual address (unicast)
                .... ..0. .... .... .... .... = LG bit: Globally unique address (factory
default)
        Source: HuaweiTe_ff:1f:a6 (54:89:98:ff:1f:a6)
            Address: HuaweiTe_ff:1f:a6 (54:89:98:ff:1f:a6)
                .... ...0 .... .... .... .... = IG bit: Individual address (unicast)
                .... ..0. .... .... .... .... = LG bit: Globally unique address (factory
default)
        Type: ARP (0x0806)
        Trailer: 000000000000000000000000000000000000
    Address Resolution Protocol (reply)
```

ARP 单播帧未到达 LSW3 前，LSW3 的交换表如下：

```
[LSW3]display mac-address
MAC address table of slot 0:
-------------------------------------------------------------------------
MAC Address    VLAN/ PEVLAN CEVLAN Port    Type    LSP/LSR-ID
VSI/SI                                     MAC-Tunnel
-------------------------------------------------------------------------
5489-98ff-1fa6 10     -      -     Eth0/0/4  dynamic  0/-
5489-9814-5d29 10     -      -     Eth0/0/2  dynamic  0/-
//目的地址为 PC1 的记录。
-------------------------------------------------------------------------
Total matching items on slot 0 displayed = 2
```

ARP 单播帧未到达 LSW1 前，LSW1 的交换表如下：

```
[LSW1]display mac-address
MAC address table of slot 0:
-------------------------------------------------------------------------
MAC Address    VLAN/ PEVLAN CEVLAN Port    Type    LSP/LSR-ID
VSI/SI                                     MAC-Tunnel
```

```
--------------------------------------------------------------------
5489-9814-5d29 10      -       -       Eth0/0/1    dynamic   0/-
//目的地址为 PC1 的记录。
5489-98ff-1fa6 10      -       -       Eth0/0/2    dynamic   0/-
--------------------------------------------------------------------
Total matching items on slot 0 displayed = 2
```

可以看到，由于单播帧从 LSW3 的 VLAN 10 进入，因此，各交换机都查找各自 VLAN 10 的交换表，并按照交换表转发。ARP 单播帧被 LSW3 转发到 LSW1，接着被 LSW1 转发到 LSW2，最后被转发到 PC1。在此过程中，其他在 VLAN 10 和 VLAN 20 内的主机都收不到该单播帧。

（5）VLAN 10 向 VLAN 20 发送的单播帧。这里由 PC1 向 PC4 发送单播帧。为了得到 PC4 的 MAC 地址，从而便于封装 PC1 ping PC4 的单播帧，因此执行以下命令，即将 LSW3 的 Ethernet0/0/2 接口先改为属于 VLAN 10：

```
[LSW3]int eth 0/0/2
[LSW3-Ethernet0/0/2]undo port default vlan
[LSW3-Ethernet0/0/2]port default vlan 10
```

执行 PC1 ping PC4 的命令，由于 PC4 现在属于 VLAN 10，因此可以 ping 通，PC1 将获得 PC4 的 MAC 地址，该 MAC 地址被缓存在 PC1 的 ARP 缓存中，便于下次需要时封装。如此，当再次执行 PC1 ping PC4 的命令时，就可以封装为一个目的地址为 PC4 的单播帧。

在 LSW3 中执行以下命令，即将 LSW3 的 Ethernet0/0/2 接口再改回属于 VLAN 20，并清空交换表：

```
[LSW3-Ethernet0/0/2]undo port default vlan
[LSW3-Ethernet0/0/2]port default vlan 20
[LSW3]undo mac-address
```

再次执行 PC1 ping PC4 的命令，可以看到，PC1 处已封装了目的 MAC 地址为 PC4 地址的 MAC 帧，如下所示：

```
No. Time       Source       Destination    Protocol    Info
1 0.000000   192.168.1.2   192.168.1.5     ICMP      Echo (ping) request (id=0x4b2d,
seq(be/le)=1/256, ttl=128)


Frame 1: 74 bytes on wire (592 bits), 74 bytes captured (592 bits)
Ethernet II, Src: HuaweiTe_14:5d:29 (54:89:98:14:5d:29), Dst: HuaweiTe_4b:23:b5
(54:89:98:4b:23:b5)
    Destination: HuaweiTe_4b:23:b5 (54:89:98:4b:23:b5)    //PC4 的 MAC 地址。
        Address: HuaweiTe_4b:23:b5 (54:89:98:4b:23:b5)
```

```
                .... ...0 .... .... .... .... = IG bit: Individual address (unicast)
                .... ..0. .... .... .... .... = LG bit: Globally unique address (factory
default)
    Source: HuaweiTe_14:5d:29 (54:89:98:14:5d:29)
        Address: HuaweiTe_14:5d:29 (54:89:98:14:5d:29)
                .... ...0 .... .... .... .... = IG bit: Individual address (unicast)
                .... ..0. .... .... .... .... = LG bit: Globally unique address (factory
default)
        Type: IP (0x0800)
    Internet Protocol, Src: 192.168.1.2 (192.168.1.2), Dst: 192.168.1.5
(192.168.1.5)
```

使用 Wireshark 观察 ICMP 协议，由于各交换机的交换表中没有对应的记录，因此该帧被交换机在 VLAN 10 中广播。显然，所有收到该帧的主机都会将其丢弃，而 PC4 则无法收到该帧。PC3 收到该帧后将其丢弃，如图 3-24 所示。

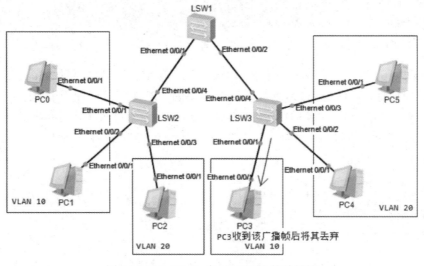

图 3-24　VLAN 10 向 VLAN 20 的单播帧

（6）验证 Trunk 链路控制。在 LSW2 中执行如下命令：

```
[LSW2]intethernet0/0/4
[LSW2-Ethernet0/0/4]undo port trunk allow-pass vlan 10
//将 VLAN 10 从 Trunk 中移除，VLAN 10 的帧无法从 Ethernet0/0/4 接口通过。
```

此时，由 PC1 ping PC3，结果是不通的。
继续执行如下命令：

```
[LSW2-Ethernet0/0/4]port trunk allow-pass vlan 10
//将 VLAN 10 添加到 Trunk 中，VLAN 10 的帧可以从 Ethernet0/0/4 接口通过。
```

再由 PC1 ping PC3，结果可以 ping 通。

读者可自行练习其他命令，增加理解。

一个 VLAN 就是一个广播域，所以在同一个 VLAN 内部，计算机之间的通信就是二层通信。如果计算机与目的计算机处在不同的 VLAN 中，那么它们之间是无法进行二层通信的，只能通过三层通信来传递信息，我们将在后面的实验中解决这个问题。

实验六：生成树配置

1. 实验目的

（1）理解生成树协议的目的和作用。
（2）掌握配置生成树协议的方法。
（3）掌握调整生成树协议中交换机的优先级的方法。

2. 生成树协议基础知识

生成树协议（Spanning Tree Protocol，STP）主要用来解决交换网络中的环路问题，即当同一个广播域中物理链路上形成环路时，其在逻辑上无法形成环路，从而避免大量广播风暴的形成。另外，生成树协议还可以为交换网络提供冗余备份链路，该协议将交换网络中的冗余备份链路从逻辑上断开，当主链路出现故障时，能够自动切换到备份链路，保证数据的正常转发。

生成树协议版本包括 STP、RSTP（快速生成树协议）和 MSTP（多生成树协议）。

生成树协议的缺点是收敛时间长。

快速生成树协议在生成树协议的基础上增加了两种接口角色——替换接口和备份接口，分别作为根接口和指定接口。当根接口或指定接口出现故障时，冗余接口可以直接切换到替换接口或备份接口，从而实现快速生成树协议小于 1s 的快速收敛。

本实验配置命令如表 3-4 所示。

表 3-4 本实验配置命令

命令格式	命令说明
display stp brief	查看当前生成树协议信息
stp priority 优先权值	设置设备 VLAN 1 的优先级，其值为 4096 的倍数，数字越小，优先级越高
stp root primary	将设备调整为 VLAN 1 的根桥

3. 实验流程

本实验观察并分析 STP 的信息，并调整设备优先级，使拓扑更为合理。实验流程如图 3-25 所示。

图 3-25 实验流程图

4. 实验步骤

（1）布置拓扑。如图 3-26 所示，拓扑中包含三台交换机 LSW0、LSW1 和 LSW2，交换机所有接口均属于 VLAN 1。在同一个广播域中，由于在物理上形成了环路，因此华为交换机默认是打开多生成树协议的，在多生成树协议的作用下，LSW1 的 Ethernet0/0/2 接口被阻塞，不能进行转发。

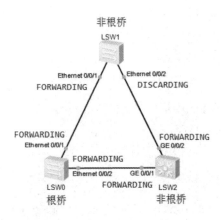

图 3-26 实验拓扑图

（2）查看交换机的 STP 信息。交换机 LSW0 的 STP 信息如下：

```
[LSW0]display stp brief
 MSTID   Port              Role  STP State    Protection
   0     Ethernet0/0/1     DESI  FORWARDING   NONE
   0     Ethernet0/0/2     DESI  FORWARDING   NONE
```

可以看出，LSW0 就是当前 VLAN 1 广播域中的根桥，其两个接口均处于转发状态。

交换机 LSW1 的 STP 信息如下：

```
[LSW1]display stp brief
 MSTID   Port              Role  STP State    Protection
   0     Ethernet0/0/1     ROOT  FORWARDING   NONE
   0     Ethernet0/0/2     ALTE  DISCARDING   NONE
```

可以看出，LSW1 不是当前 VLAN 1 广播域中的根桥，其 Ethernet0/0/1 接口是根接口，通往根桥，Ethernet0/0/2 接口被阻塞，这样就形成一种逻辑上的树形结构，防止了环路。

交换机 LSW2 的 STP 信息如下：

```
[LSW2]display stp brief
 MSTID   Port                      Role  STP State    Protection
   0     GigabitEthernet0/0/1      ROOT  FORWARDING   NONE
   0     GigabitEthernet0/0/2      DESI  FORWARDING   NONE
```

显然，该三层交换机不是根桥，其 Ethernet0/0/1 接口是根接口，通往根桥，两个接口都处于转发状态。若将 Ethernet0/0/1 接口 shutdown，则 Ethernet0/0/2 接口将从 DISCARDING 状态切换到 FORWARDING 状态。这样，网络的实际拓扑就变成如图 3-27 所示的结构。

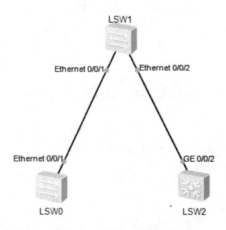

图 3-27　切换后的实际拓扑

（3）调整优先级，使三层交换机 LSW2 成为根桥。在 LSW2 中做如下配置，指定三层交换机为 VLAN 1 的根桥：

```
[LSW2]stp instance 0 root primary
```

执行上述命令后，再次查看 STP 信息：

```
[LSW2]display stp brief
MSTID  Port                    Role  STP State    Protection
0      GigabitEthernet0/0/1    DESI  FORWARDING   NONE
0      GigabitEthernet0/0/2    DESI  FORWARDING   NONE
```

通过对比可以发现，LSW2 已经成为根桥，其优先级数字变小，意味着优先级提高了；同时两个接口都变为 FORWARDING 状态。LSW2 为根桥后的实际拓扑如图 3-28 所示。通过命令行可以看到，LSW1 的 Ethernet0/0/1 接口变为阻塞状态，交换机 LSW1 的 STP 信息如下：

```
[LSW1]dis stp brief
MSTID  Port            Role  STP State    Protection
0      Ethernet0/0/1   ALTE  DISCARDING   NONE      //变为阻塞状态。
0      Ethernet0/0/2   ROOT  FORWARDING   NONE
```

也可以通过直接改变优先级数字来达到目的。比如，在 LSW0 中执行如下命令也可将 LSW0 改为根桥：

```
[LSW0]stp instance 0 priority 4096
```

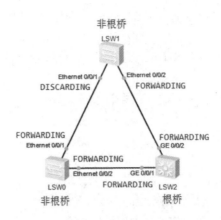

图 3-28 LSW2 为根桥后的实际拓扑

实验七：链路聚合配置

1. 实验目的

（1）理解链路聚合的目的和作用。
（2）掌握链路聚合的要求和条件。
（3）掌握链路聚合的配置。

2. 链路聚合基础知识

链路聚合（Eth-Trunk）是指交换机将多个物理接口聚合成一个逻辑接口，可将其理解为一个接口。通过链路聚合，可以提高交换机间的带宽。例如，当 2 个 100Mb/s 带宽的接口聚合后，就可以生成一个 200Mb/s 带宽的逻辑接口。在某种情况下，当带宽不够而又有多余接口时，可以通过聚合来满足需求，节省费用。

一个链路聚合内的几个物理接口还可以实现负载均衡，当某个接口出现故障时，逻辑接口内的其他接口将自动承载其余的流量。

参与聚合的各接口必须具有相同的属性，如速率、Trunk 模式、单/双工模式等。

链路聚合可以使用手工方式配置，也可以使用动态协议来配置。

本实验配置命令如表 3-5 所示。

表 3-5 本实验配置命令

命令格式	命令说明
interface eth-trunk 聚合逻辑接口号	用来在系统模式下创建聚合接口号，如[Huawei]int eth-trunk 1，该命令创建聚合逻辑接口号 1
trunkport GigabitEthernet/Ethernet 物理接口号	用来在接口模式下应用聚合接口
mode lacp-static/manual	用来配置链路聚合模式
load-balance 负载平衡方式	可按源 IP 地址、目的 IP 地址、源 MAC 地址、目的 MAC 地址进行负载平衡

续表

命令格式	命令说明
display eth-trunk trunk ID	用来查看特定的链路聚合状态
display eth-trunk	用来查看链路聚合汇总信息

3. 实验流程

本实验配置以太通道，将三个 100Mb/s 带宽的物理接口聚合为一条 300Mb/s 带宽的以太通道。实验流程如图 3-29 所示。

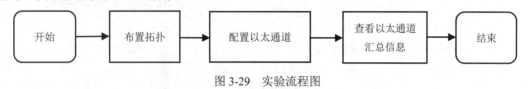

图 3-29　实验流程图

4. 实验步骤

（1）布置拓扑。如图 3-30 所示，拓扑中两台交换机的 Ethernet0/0/1、Ethernet0/0/2 和 Ethernet0/0/3 三个接口分别对应连接，但只有一条链路是通的，这是因为生成树默认开启，另两条链路被阻塞了。

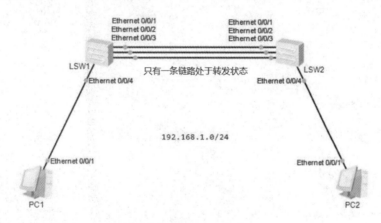

图 3-30　实验拓扑图

在交换机 LSW1 上查看 STP 接口堵塞情况：

```
[LSW1]display stp brief
 MSTID   Port            Role    STP State    Protection
    0    Ethernet0/0/1   DESI    FORWARDING   NONE
    0    Ethernet0/0/2   DESI    FORWARDING   NONE
    0    Ethernet0/0/3   DESI    FORWARDING   NONE
    0    Ethernet0/0/4   DESI    FORWARDING   NONE
```

在交换机 LSW2 上查看 STP 接口堵塞情况：

```
[LSW2]display stp brief
 MSTID   Port                Role   STP State    Protection
    0    Ethernet0/0/1       ROOT   FORWARDING   NONE
    0    Ethernet0/0/2       ALTE   DISCARDING   NONE
    0    Ethernet0/0/3       ALTE   DISCARDING   NONE
    0    Ethernet0/0/4       DESI   FORWARDING   NONE
```

（2）配置以太通道。通过配置以太通道，使连接交换机的三条链路全部起作用，如图 3-31 所示。

图 3-31 配置以太通道后链路全部开启

交换机 LSW1：

```
<LSW1>sy
Enter system view, return user view with Ctrl+Z.
[LSW1]interface Eth-Trunk 1
//创建链路聚合1，通道范围为 0～63。
[LSW1-Eth-Trunk1]trunkport Ethernet 0/0/1 to 0/0/3
//将三个物理接口加入链路聚合1中。
[LSW1-Eth-Trunk1]load-balance ?
//下面为负载均衡可选项。
  dst-ip       According to destination IP hash arithmetic
  dst-mac      According to destination MAC hash arithmetic
  src-dst-ip   According to source/destination IP hash arithmetic
  src-dst-mac  According to source/destination MAC hash arithmetic
  src-ip       According to source IP hash arithmetic
  src-mac      According to source MAC hash arithmetic
[LSW1-Eth-Trunk1]load-balance src-mac
//选择按源 MAC 地址负载均衡。
[LSW1-Eth-Trunk1]port link-type trunk
//将链路聚合设为 Trunk 模式。
```

交换机 LSW2：

```
<LSW1>sy
Enter system view, return user view with Ctrl+Z.
[LSW2]int eth-trunk 1
[LSW2-Eth-Trunk1]trunkport Ethernet 0/0/1 to 0/0/3
[LSW2-Eth-Trunk1]load-balance src-mac
[LSW2-Eth-Trunk1]port link-type trunk
```

在交换机 LSW1 上查看 STP 接口堵塞情况：

```
[LSW1]display stp brief
 MSTID    Port              Role     STP State      Protection
    0     Ethernet0/0/4     DESI     FORWARDING     NONE
    0     Eth-Trunk1        DESI     FORWARDING     NONE
```

在交换机 LSW2 上查看 STP 接口堵塞情况：

```
[LSW2]display stp brief
 MSTID    Port              Role     STP State      Protection
    0     Ethernet0/0/4     DESI     FORWARDING     NONE
    0     Eth-Trunk1        ROOT     FORWARDING     NONE
```

（3）验证两台主机能否 ping 通。读者可自行练习。
（4）查看链路聚合的汇总信息。

交换机 LSW1：

```
[LSW1]display eth-trunk 1
Eth-Trunk1's state information is:
WorkingMode: NORMAL          Hash arithmetic: According to SA
Least Active-linknumber: 1   Max Bandwidth-affected-linknumber: 8
Operate status: up           Number Of Up Port In Trunk: 3
--------------------------------------------------------------------
PortName                  Status          Weight
Ethernet0/0/1             Up              1
Ethernet0/0/2             Up              1
Ethernet0/0/3             Up              1
```

交换机 LSW2：

```
[LSW2]display eth-trunk 1
Eth-Trunk1's state information is:
```

```
WorkingMode: NORMAL          Hash arithmetic: According to SA
Least Active-linknumber: 1  Max Bandwidth-affected-linknumber: 8
Operate status: up           Number Of Up Port In Trunk: 3
--------------------------------------------------------------------
PortName                     Status      Weight
Ethernet0/0/1                Up          1
Ethernet0/0/2                Up          1
Ethernet0/0/3                Up          1
```

第 4 章   网络层

实验一：路由器 IP 地址配置及直连网络

1. 实验目的

（1）理解 IP 地址。
（2）掌握路由器接口 IP 地址的配置方法。
（3）理解路由器的直连网络。

2. 基础知识

IP 地址是网络层中使用的地址，不管网络层下面是什么网络或什么类型的接口，在网络层看来，它只是一个可以用 IP 地址代表的接口地址而已。网络层依靠 IP 地址和路由协议将数据报送到目的 IP 地址。既然是一个地址，那么一个 IP 地址就只能代表一个接口，否则会造成地址的二义性；接口则不同，一个接口可以配多个 IP 地址，这并不会造成地址的二义性。

路由器是互联网的核心设备，它在 IP 网络间转发数据报，因此路由器的每个接口都连接一个或多个网络，而两个接口却不可以代表同一个网络。路由器的一个配置了 IP 地址的接口所在的网络就是路由器的直连网络。对于直连网络，路由器并不需要额外对其配置路由，当其接口被激活后，路由器会自动将直连网络加入路由表中。

本实验配置命令如表 4-1 所示。

表 4-1 本实验配置命令

命令格式	命令说明
ip address IP 地址 子网掩码	接口模式下给当前接口配置 IP 地址，例如：ip address 192.168.1.1 255.255.255.0
display ip routing-table	查看路由器的路由表
display ip interface brief	查看路由器的 IP 地址配置情况
undo shutdown	接口模式下激活当前接口（华为路由器接口默认开启）

3. 实验流程

实验流程如图 4-1 所示。

4. 实验步骤

（1）布置拓扑。如图 4-2 所示，路由器连接了两个网络，通过 GE0/0/0 接口连接网络 192.168.1.0/24，通过 GE0/0/1 接口连接网络 192.168.2.0/24，这两个网络都属于路由器的直连网络。

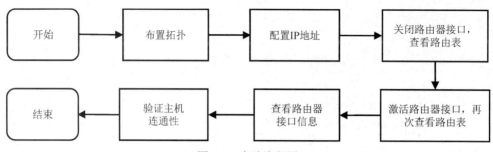

图 4-1　实验流程图

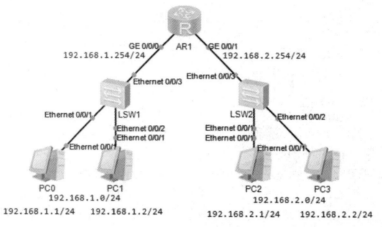

图 4-2　实验拓扑图

（2）配置路由器的 IP 地址：

```
<AR1>sy
Enter system view, return user view with Ctrl+Z.
[AR1]int g0/0/0
[AR1-GigabitEthernet0/0/0]ip address 192.168.1.254 24
[AR1-GigabitEthernet0/0/0]shutdown
//华为路由器接口默认开启，为便于观察，先将接口关闭。
[AR1-GigabitEthernet0/0/0]quit
[AR1]int g0/0/1
[AR1-GigabitEthernet0/0/1]ip address 192.168.2.254 24
[AR1-GigabitEthernet0/0/1]shutdown
[AR1-GigabitEthernet0/0/1]quit
```

（3）查看路由表，结果如下：

```
[AR1]display ip routing-table
//查看路由表，可以看到路由表是空的。路由表中以"127"开头的路由是本地环回地址形成的路由，
//255.255.255.255 是广播地址。
Route Flags: R - relay, D - download to fib
```

```
------------------------------------------------------------
Routing Tables: Public
        Destinations : 4       Routes : 4
Destination/Mask    Proto  Pre  Cost    Flags  NextHop    Interface
      127.0.0.0/8   Direct  0    0        D    127.0.0.1  InLoopBack0
      127.0.0.1/32  Direct  0    0        D    127.0.0.1  InLoopBack0
127.255.255.255/32  Direct  0    0        D    127.0.0.1  InLoopBack0
255.255.255.255/32  Direct  0    0        D    127.0.0.1  InLoopBack0
```

（4）激活接口：

```
[AR1]int g0/0/0
[AR1-GigabitEthernet0/0/0]undo shutdown        //激活接口。
[AR1-GigabitEthernet0/0/0]quit
[AR1]int g0/0/1
[AR1-GigabitEthernet0/0/1]undo shutdown
[AR1-GigabitEthernet0/0/1]quit
```

（5）查看路由表，观察路由表的变化（注意 Proto 为 Direct 的路由条目是直连路由）。结果如下：

```
[AR1]display ip routing-table            //查看路由表。
Route Flags: R - relay, D - download to fib
------------------------------------------------------------
Routing Tables: Public
        Destinations : 10      Routes : 10
Destination/Mask    Proto  Pre  Cost  Flags  NextHop        Interface
      127.0.0.0/8   Direct  0    0     D    127.0.0.1      InLoopBack0
      127.0.0.1/32  Direct  0    0     D    127.0.0.1      InLoopBack0
127.255.255.255/32  Direct  0    0     D    127.0.0.1      InLoopBack0
    192.168.1.0/24  Direct  0    0     D    192.168.1.254  GigabitEthernet0/0/0
//直连路由。
  192.168.1.254/32  Direct  0    0     D    127.0.0.1      GigabitEthernet0/0/0
//路由器自身的 IP 地址。
  192.168.1.255/32  Direct  0    0     D    127.0.0.1      GigabitEthernet0/0/0
    192.168.2.0/24  Direct  0    0     D    192.168.2.254  GigabitEthernet0/0/1
  192.168.2.254/32  Direct  0    0     D    127.0.0.1      GigabitEthernet0/0/1
  192.168.2.255/32  Direct  0    0     D    127.0.0.1      GigabitEthernet0/0/1
255.255.255.255/32  Direct  0    0     D    127.0.0.1      InLoopBack0
```

（6）查看接口信息，结果如下：

```
[AR1]display interface GigabitEthernet 0/0/0          //查看接口信息。
GigabitEthernet0/0/0 current state : UP
Line protocol current state : UP
Last line protocol up time : 2020-10-13 14:16:09 UTC-08:00
Description:HUAWEI, AR Series, GigabitEthernet0/0/0 Interface
Route Port,The Maximum Transmit Unit is 1500
Internet Address is 192.168.1.254/24
IP Sending Frames' Format is PKTFMT_ETHNT_2, Hardware address is 00e0-fc32-1c5e
Last physical up time    : 2020-10-13 14:16:09 UTC-08:00
Last physical down time : 2020-10-13 14:13:11 UTC-08:00
Current system time: 2020-10-13 14:22:54-08:00
Port Mode: FORCE COPPER
Speed : 1000,  Loopback: NONE
Duplex: FULL,  Negotiation: ENABLE
Mdi   : AUTO
Last 300 seconds input rate 424 bits/sec, 0 packets/sec
Last 300 seconds output rate 0 bits/sec, 0 packets/sec
Input peak rate 568 bits/sec,Record time: 2020-10-13 14:04:11
Output peak rate 96 bits/sec,Record time: 2020-10-13 14:10:56

Input:  431 packets, 51289 bytes
  Unicast:                   0,  Multicast:                 431
  Broadcast:                 0,  Jumbo:                       0
  Discard:                   0,  Total Error:                 0

  CRC:                       0,  Giants:                      0
  Jabbers:                   0,  Throttles:                   0
  Runts:                     0,  Symbols:                     0
  Ignoreds:                  0,  Frames:                      0

Output:  2 packets, 120 bytes
  Unicast:                   0,  Multicast:                   0
  Broadcast:                 2,  Jumbo:                       0
  Discard:                   0,  Total Error:                 0

  Collisions:                0,  ExcessiveCollisions:         0
  Late Collisions:           0,  Deferreds:                   0
```

```
        Input bandwidth utilization threshold : 100.00%
        Output bandwidth utilization threshold: 100.00%
        Input bandwidth utilization  :    0%
        Output bandwidth utilization :    0%
```

（7）验证连通性。从主机端使用 ping 命令来测试网络的连通性。

另外，若给 GE0/0/1 接口配置的 IP 地址为 192.168.1.3/24，则会弹出出错信息提示，表示该指定的 IP 地址与另一个 IP 地址冲突，如图 4-3 所示。也就是说，不同路由器接口所连接的不能是同一个网络。

```
[AR1-GigabitEthernet0/0/1]ip address 192.168.1.3 24
Error: The specified address conflicts with another address.
```

图 4-3　IP 地址冲突提示

实验二：ARP 协议分析

1. 实验目的

（1）理解 ARP 协议的作用。
（2）理解 ARP 协议的工作方式。

2. 基础知识

互联网常被解释为"网络的网络"，其思想是把各种网络都统一到一个网络中来，并采用一种统一的地址（IP 地址）使各种网络在路由协议的作用下实现互联。但这里面有一个重要问题，即互联网是基于 IP 网络去路由的，而被互联网连接起来的其他网络，比如以太网，它们内部是使用自己的 MAC 地址去寻址的，当到达一个以太网的网段时，就需要知道目的 IP 地址对应的 MAC 地址，这样，才能最终将 IP 数据报送到目的 IP 地址。实际上，这样的过程一直存在。

ARP 协议用来解决局域网内一个广播域中的 IP 地址和 MAC 地址的映射问题。其中，ARP 请求是广播分组，该广播域内的主机都可以收到，ARP 响应是单播分组，由响应主机直接发给请求主机，详细解释请参见教材《计算机网络（第 8 版）》4.2.4 节。ARP 分组的格式如图 4-4 所示。

为了提高效率，避免 ARP 请求占用过多的网络资源，主机或路由器都设置有 ARP 高速缓存，用来将请求得到的映射保存起来，以备下次需要时直接使用。ARP 高速缓存设有时间限制，防止因地址改变且没有及时更新造成发送失败的情况。

当然，如果源主机本身发送的就是广播分组，或双方使用的是点对点的链路，就不需要发起 ARP 请求了。

看下面的例子。如图 4-5 所示，两台主机由三台路由器连接，且接口均使用快速以太网接口，则由 PC0 到 PC1 的分组发送过程中共经历了四次 ARP 请求。在此过程中，源 IP 地址和目的 IP 地址是始终不变的,而源 MAC 地址和目的 MAC 地址在不同的二层广播域中会改变。

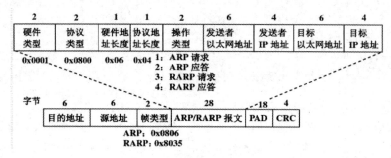

图 4-4 ARP 分组的格式

图 4-5 ARP 请求

3. 实验流程

实验流程如图 4-6 所示。

图 4-6 实验流程图

4. 实验步骤

（1）布置拓扑。如图 4-7 所示，路由器连接了两个网络，通过 GE0/0/0 接口连接网络 192.168.1.0/24，通过 GE0/0/1 接口连接网络 192.168.2.0/24，这两个网络都属于路由器的直连网络。

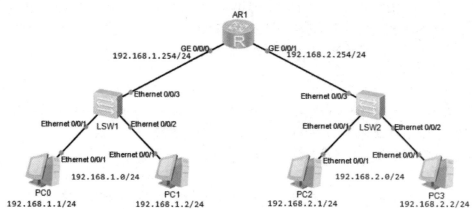

图 4-7 实验拓扑图

（2）由 PC0 ping PC3，在 LSW1 的 Ethernet0/0/3 接口使用 Wireshark 抓取数据包，观察 ARP 分组的走向及结构。由于目的 IP 地址和源 IP 地址不在同一网络，因此 PC0 应将 IP 分组发送给自己的网关，即路由器。这样，PC0 需要通过 ARP 请求分组得到网关的 MAC 地址，用于发往网关的数据链路层封装。当 PC0 得到网关的 MAC 地址后，会将其添加到自己的 ARP 高速缓存中，当在生存期内再次访问网关时，就不需要发出对网关的 ARP 请求了。

此处 PC0 生成 ARP 请求分组，该请求分组将通过交换机被广播到 PC1 和 AR1。PC1 会将其丢弃，只有 AR1 会收下该请求分组，并做出响应，如图 4-8 所示。

ARP 分组的走向及结构如下：

```
No.  Time        Source              Destination        Protocol        Info
1    0.000000    HuaweiTe_5f:58:ba   Broadcast          ARP             Who has 192.168.1.254? Tell 192.168.1.1

Ethernet II, Src: HuaweiTe_5f:58:ba (54:89:98:5f:58:ba), Dst: Broadcast (ff:ff:ff:ff:ff:ff)
    Destination: Broadcast (ff:ff:ff:ff:ff:ff)
        Address: Broadcast (ff:ff:ff:ff:ff:ff)
        .... ...1 .... .... .... .... = IG bit: Group address (multicast/broadcast)
        .... ..1. .... .... .... .... = LG bit: Locally administered address (this is NOT the factory default)
    Source: HuaweiTe_5f:58:ba (54:89:98:5f:58:ba)
        Address: HuaweiTe_5f:58:ba (54:89:98:5f:58:ba)
        .... ...0 .... .... .... .... = IG bit: Individual address (unicast)
        .... ..0. .... .... .... .... = LG bit: Globally unique address (factory default)
    Type: ARP (0x0806)
    Trailer: 000000000000000000000000000000000000
```

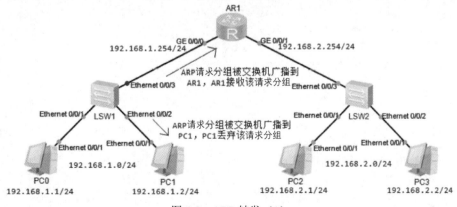

图 4-8 ARP 转发（1）

AR1 收下 ARP 请求分组，将 PC0 的 IP 地址和 MAC 地址记入 ARP 高速缓存，并生成

ARP 的响应分组，然后把该响应分组以单播的形式发送给 PC0，结果如下：

```
No.  Time       Source              Destination         Protocol      Info
  2  0.031000   HuaweiTe_4c:7f:14   HuaweiTe_5f:58:ba   ARP           192.168.1.254
is at 00:e0:fc:4c:7f:14

Ethernet II, Src: HuaweiTe_4c:7f:14 (00:e0:fc:4c:7f:14), Dst:
HuaweiTe_5f:58:ba (54:89:98:5f:58:ba)
    Destination: HuaweiTe_5f:58:ba (54:89:98:5f:58:ba)
        Address: HuaweiTe_5f:58:ba (54:89:98:5f:58:ba)
        .... ...0 .... .... .... .... = IG bit: Individual address (unicast)
        .... ..0. .... .... .... .... = LG bit: Globally unique address (factory
default)
    Source: HuaweiTe_4c:7f:14 (00:e0:fc:4c:7f:14)
        Address: HuaweiTe_4c:7f:14 (00:e0:fc:4c:7f:14)
        .... ...0 .... .... .... .... = IG bit: Individual address (unicast)
        .... ..0. .... .... .... .... = LG bit: Globally unique address (factory
default)
    Type: ARP (0x0806)
    Trailer: 000000000000000000000000000000000000
```

PC0 收到该响应分组后，就得到了网关（192.168.1.254）的 MAC 地址。PC0 会封装网关的 MAC 地址，并发送给网关，即 AR1 的 GE0/0/0 接口。而 AR1 会查询路由表，ARP 请求分组将从 GE0/0/1 接口被转发出去，故在 GE0/0/1 接口处封装 MAC 帧时，就需要目的 IP 地址 192.168.2.2 的 MAC 地址。由于是第一次，AR1 的缓存中并没有保存该 IP 地址对应的 MAC 地址，因此需要发出 ARP 请求分组来获得需要的 MAC 地址，如图 4-9 和图 4-10 所示。观察该请求分组的广播域。

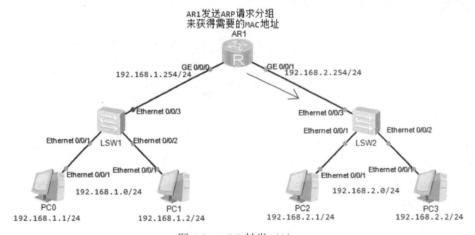

图 4-9 ARP 转发（2）

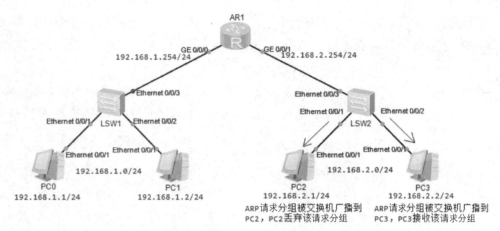

图4-10 ARP 转发（3）

PC3 收到的 ARP 分组如下：

```
No.    Time       Source              Destination    Protocol    Info
 1  0.000000   HuaweiTe_4c:7f:15      Broadcast      ARP    Who has 192.168.2.2? Tell 192.168.2.254

Ethernet II, Src: HuaweiTe_4c:7f:15 (00:e0:fc:4c:7f:15), Dst: Broadcast (ff:ff:ff:ff:ff:ff)
    Destination: Broadcast (ff:ff:ff:ff:ff:ff)
        Address: Broadcast (ff:ff:ff:ff:ff:ff)
        .... ...1 .... .... .... .... = IG bit: Group address (multicast/broadcast)
        .... ..1. .... .... .... .... = LG bit: Locally administered address (this is NOT the factory default)
    Source: HuaweiTe_4c:7f:15 (00:e0:fc:4c:7f:15)
        Address: HuaweiTe_4c:7f:15 (00:e0:fc:4c:7f:15)
        .... ...0 .... .... .... .... = IG bit: Individual address (unicast)
        .... ..0. .... .... .... .... = LG bit: Globally unique address (factory default)
    Type: ARP (0x0806)
    Trailer: 000000000000000000000000000000000000
```

PC3 封装的 ARP 分组如下：

```
No.    Time       Source              Destination         Protocol    Info
 2  0.031000   HuaweiTe_e9:5b:dd      HuaweiTe_4c:7f:15   ARP    192.168.2.2 is at 54:89:98:e9:5b:dd

Ethernet II, Src: HuaweiTe_e9:5b:dd (54:89:98:e9:5b:dd), Dst:
```

```
HuaweiTe_4c:7f:15 (00:e0:fc:4c:7f:15)
    Destination: HuaweiTe_4c:7f:15 (00:e0:fc:4c:7f:15)
        Address: HuaweiTe_4c:7f:15 (00:e0:fc:4c:7f:15)
        .... ...0 .... .... .... .... = IG bit: Individual address (unicast)
        .... ..0. .... .... .... .... = LG bit: Globally unique address (factory
default)
    Source: HuaweiTe_e9:5b:dd (54:89:98:e9:5b:dd)
        Address: HuaweiTe_e9:5b:dd (54:89:98:e9:5b:dd)
        .... ...0 .... .... .... .... = IG bit: Individual address (unicast)
        .... ..0. .... .... .... .... = LG bit: Globally unique address (factory
default)
    Type: ARP (0x0806)
    Trailer: 00000000000000000000000000000000000000
```

实验三：静态路由与默认路由配置

1. 实验目的

（1）理解静态路由的含义。
（2）掌握路由器静态路由的配置方法。
（3）理解默认路由的含义。
（4）掌握路由器默认路由的配置方法。

2. 基础知识

静态路由是指由网络信息由网络管理员手工配置，而不是路由器通过路由算法和其他路由器学习得到的。所以，静态路由主要适合网络规模不大、拓扑结构相对固定的网络。当网络环境比较复杂时，由于其拓扑或链路状态容易变化，因此需要网络管理员手工改变路由，这对网络管理员来说是一个烦琐的工作，且网络容易受人的影响，对网络管理员不论技术上还是纪律上都有更高的要求。

默认路由也是一种静态路由，它位于路由表的最后，当数据报与路由表中前面的表项都不匹配时，数据报将根据默认路由转发。这种机制在某些时候是非常有效的，如在末梢网络中，默认路由可以大大简化路由器的项目数量及配置，减轻路由器和网络管理员的工作负担。可见，静态路由优先级高于默认路由。

本实验配置命令如表 4-2 所示。

表 4-2　本实验配置命令

命令格式	命令说明
[AR]ip route-static 目的网络号 目的网络掩码 下一跳 IP 地址	配置静态路由
[AR]ip route-static 0.0.0.0 0.0.0.0 下一跳 IP 地址	配置默认路由

3. 实验流程

本实验配置静态路由和默认路由,要求各 IP 地址全部可达。实验流程如图 4-11 所示。

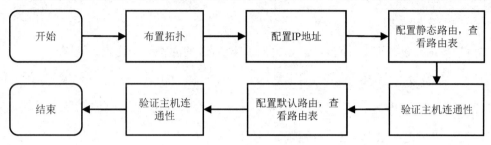

图 4-11 实验流程图

4. 实验步骤

(1) 按图 4-12 所示布置拓扑,并按表 4-3 配置各设备的 IP 地址。

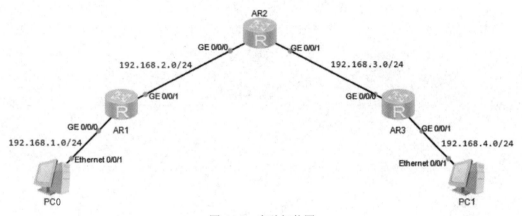

图 4-12 实验拓扑图

表 4-3 各设备的 IP 地址配置

设备名称	接口	IP 地址	默认网关
AR1	GE0/0/0	192.168.1.254/24	
	GE0/0/1	192.168.2.1/24	
AR2	GE0/0/0	192.168.2.2/24	
	GE0/0/1	192.168.3.1/24	
AR3	GE0/0/0	192.168.3.2/24	
	GE0/0/1	192.168.4.254/24	
PC0	Ethernet0/0/1	192.168.1.1/24	192.168.1.254
PC1	Ethernet0/0/1	192.168.4.1/24	192.168.4.254

(2) 配置静态路由。

首先配置路由器 AR1 的路由。AR1 有两个直连网络,分别是 192.168.1.0/24 和 192.168.2.0/24,这两个网络不需要配置静态路由。由于 AR1 不知道 192.168.3.0/24 和 192.168.4.0/24 这两个网络的路由,因此需要在 AR1 上配置这两个静态路由,这需要网络管理员人工判断下一跳地址。配置如下：

```
[AR1]ip route-static 192.168.3.0 24 192.168.2.2
[AR1]ip route-static 192.168.4.0 24 192.168.2.2
```

然后配置路由器 AR2 的路由。同理可分析出 AR2 的静态路由配置，配置如下：

```
[AR2]ip route-static 192.168.1.0 24 192.168.2.1
[AR2]ip route-static 192.168.4.0 24 192.168.3.2
```

最后配置路由器 AR3 的路由。配置如下：

```
[AR3]ip route-static 192.168.1.0 24 192.168.3.1
[AR3]ip route-static 192.168.2.0 24 192.168.3.1
```

以 AR1 为例查看路由器的路由表，其中 Proto 为 Static 的路由为静态路由，Proto 为 Direct 的路由为直连路由。AR2 和 AR3 的路由表请读者自行分析。查看 AR1 的路由表，结果如下：

```
[AR1]dis ip routing-table
Route Flags: R - relay, D - download to fib
------------------------------------------------------------------
Routing Tables: Public
         Destinations : 12       Routes : 12
   Destination/Mask    Proto  Pre  Cost Flags NextHop        Interface
        127.0.0.0/8    Direct 0    0     D    127.0.0.1      InLoopBack0
        127.0.0.1/32   Direct 0    0     D    127.0.0.1      InLoopBack0
  127.255.255.255/32   Direct 0    0     D    127.0.0.1      InLoopBack0
      192.168.1.0/24   Direct 0    0     D    192.168.1.254  GigabitEthernet0/0/0
    192.168.1.254/32   Direct 0    0     D    127.0.0.1      GigabitEthernet0/0/0
    192.168.1.255/32   Direct 0    0     D    127.0.0.1      GigabitEthernet0/0/0
      192.168.2.0/24   Direct 0    0     D    192.168.2.1    GigabitEthernet0/0/1
      192.168.2.1/32   Direct 0    0     D    127.0.0.1      GigabitEthernet0/0/1
    192.168.2.255/32   Direct 0    0     D    127.0.0.1      GigabitEthernet0/0/1
      192.168.3.0/24   Static 60   0     RD   192.168.2.2    GigabitEthernet0/0/1
      192.168.4.0/24   Static 60   0     RD   192.168.2.2    GigabitEthernet0/0/1
  255.255.255.255/32   Direct 0    0     D    127.0.0.1      InLoopBack0
```

由 PC0 ping PC1，验证是否能 ping 通。

（3）配置默认路由。AR1 有两个直连网络，分别是 192.168.1.0/24 和 192.168.2.0/24，这两个网络不需要配置静态路由。通过前面对静态路由的分析可知，AR1 去 192.168.3.0/24 和 192.168.4.0/24 这两个网络的下一跳 IP 地址都是 192.168.2.2，所以，这两个静态路由可以由一条指向 192.168.2.2 的默认路由代替。在前面配置的基础上，将静态路由删除（静态路由前面加 undo），再增加一条默认路由即可。配置如下：

```
[AR1]undo ip route-static 192.168.3.0 255.255.255.0 192.168.2.2
[AR1]undo ip route-static 192.168.4.0 255.255.255.0 192.168.2.2
[AR1]ip route-static 0.0.0.0 0.0.0.0 192.168.2.2
```

配置 AR3 可参考 AR1 的配置方法。配置如下：

```
[AR3]undo ip route-static 192.168.1.0 255.255.255.0 192.168.3.1
[AR3]undo ip route-static 192.168.2.0 255.255.255.0 192.168.3.1
[AR3]ip route-static 0.0.0.0 0.0.0.0 192.168.3.1
```

以 AR1 为例查看路由器的路由表，其中 Destination/Mask 为 0.0.0.0/0，Proto 为 Static 的路由为默认路由。查看 AR1 的路由表，结果如下：

```
[AR1]display ip routing-table
Route Flags: R - relay, D - download to fib
------------------------------------------------------------
Routing Tables: Public
         Destinations : 11      Routes : 11
Destination/Mask    Proto   Pre  Cost  Flags  NextHop          Interface
       0.0.0.0/0   Static   60   0     RD     192.168.2.2     GigabitEthernet0/0/1
     127.0.0.0/8   Direct   0    0     D      127.0.0.1       InLoopBack0
    127.0.0.1/32   Direct   0    0     D      127.0.0.1       InLoopBack0
127.255.255.255/32 Direct   0    0     D      127.0.0.1       InLoopBack0
   192.168.1.0/24  Direct   0    0     D      192.168.1.254   GigabitEthernet0/0/0
 192.168.1.254/32  Direct   0    0     D      127.0.0.1       GigabitEthernet0/0/0
 192.168.1.255/32  Direct   0    0     D      127.0.0.1       GigabitEthernet0/0/0
   192.168.2.0/24  Direct   0    0     D      192.168.2.1     GigabitEthernet0/0/1
   192.168.2.1/32  Direct   0    0     D      127.0.0.1       GigabitEthernet0/0/1
 192.168.2.255/32  Direct   0    0     D      127.0.0.1       GigabitEthernet0/0/1
255.255.255.255/32 Direct   0    0     D      127.0.0.1       InLoopBack0
```

由 PC0 ping PC1，验证是否能 ping 通。

实验四：RIP 路由协议配置

1. 实验目的

（1）理解 RIP 路由的原理。
（2）掌握 RIP 路由的配置方法。

2. 基础知识

RIP（Routing Information Protocol）属于内部网关协议（IGP），用于一个自治系统内部。RIP 是一种基于距离向量的分布式的路由选择协议，实现简单，应用较为广泛。RIP 的中文名称为路由信息协议，但很少被提及，更多的是被更为简洁的英文简称代替。

RIP 是在 20 世纪 70 年代从美国的 Xerox 公司开发的早期协议——网关信息协议（GWINFO）中演变而来的。目前，RIP 共有三个版本，分别是 RIPv1、RIPv2 和 RIPng，其中 RIPv1 和 RIPv2 用于 IPv4 网络，RIPng 用于 IPv6 网络。由于 RIP 不支持子网及跳数太少等原因，实际上常用的是 RIPv2。可从以下几方面理解 RIP 的特点。

（1）在 RIP 中，距离最短的路由就是最好的路由。RIP 对距离的度量是跳数，初始的直连路由距离为 1，此后每经过一台路由器，跳数就加 1，这样，经过的路由器数量越多，距离也就越大。RIP 规定，一条路由最大的跳数为 15，也就是最大距离为 16，距离超出 16 的路由被认为不可达，会被删除。

（2）RIP 中路由的更新是通过定时广播实现的，接收对象为邻居。在缺省情况下，路由器每隔 30s 向与它相连的网络广播自己的路由表，接到广播的路由器将收到的信息按一定算法添加到自身的路由表中。每台路由器都这样广播，最终网络上所有的路由器都会得知全部 RIP 范围的路由信息。

（3）环路的解决办法。RIP 中也存在环路问题，如好消息传播得快，坏消息传播得慢。解决办法通常有以下几种。

① 定义最大跳数。比如，将 TTL 值设为 16，若分组陷入路由循环中，则跳数耗尽后就会被消灭，在 RIP 中就被视为网络不可达而被删除。

② 水平分割。水平分割即单向路由更新，它保证路由器记住每一条路由信息的来源，并且不在收到这条信息的接口上再次发送它，这是不产生路由循环的最基本措施。若 A 从 B 处得到一个网络的路由信息，则 A 不会向 B 更新该网络可以通过 B 到达的信息。这样，当该网络出现故障不可达时，B 会将路由信息通告给 A，而 A 则不会把可以通过 B 到达该网络的路由信息通告给 B。如此便可以加快网络收敛，破坏路由环路。

③ 路由毒化。当某直连网络发生故障时，路由器将其度量值标为无穷大，并将此路由信息通告给邻居，邻居再向其邻居通告，依次毒化各路由器，从而避免环路。

④ 控制更新时间。也称抑制计时，当一条路由信息无效之后，就在一段时间内使这条路由处于抑制状态，即不再接收关于相同目的地址的路由更新。显然，当一个网络频繁地在有效和无效之间切换时，往往是有问题的，这时，若将该网络的路由信息在一定时间内不更新，则可以增加网络的稳定性，避免路由振荡。

（4）RIPv1 和 RIPv2 的主要区别如下。

① RIPv1 是有类路由协议，RIPv2 是无类路由协议。

② RIPv1 不能支持 VLSM（变长子网掩码），RIPv2 可以支持 VLSM。

③ RIPv1 没有认证的功能，RIPv2 可以支持认证，并且有明文和 MD5 两种认证。

④ RIPv1 没有手工汇总的功能，RIPv2 可以在关闭自动汇总的前提下，进行手工汇总。

⑤ RIPv1 是广播更新，RIPv2 是组播更新。

⑥ RIPv1 对路由没有标记的功能，RIPv2 可以对路由打标记（tag），用于过滤和做策略。

详细内容请参见教材《计算机网络（第 8 版）》4.6.2 节。

本实验配置命令如表 4-4 所示。

表 4-4 本实验配置命令

命令格式	命令说明
sy name 路由器名称	配置路由器名称
rip rip-id	启动 RIP 路由协议
version 版本号	设置 RIP 版本，可为 1 或者 2
network 网络号	网络号应为路由器直连的网络号，是分类网络号
debugging rip rip-id	显示 RIP 路由的动态更新
summary	路由汇总
display rip rip-id	显示路由协议配置与统计等信息
silent-interface 接口名	将接口设置为静默接口，此接口不再发送路由信息

3. 实验流程

实验流程如图 4-13 所示。

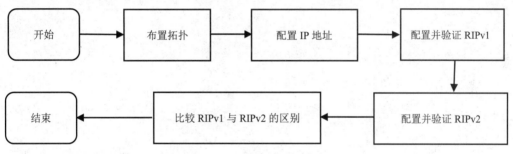

图 4-13 实验流程图

4. RIPv1 实验步骤

（1）按图 4-14 所示布置拓扑，并按表 4-5 配置各设备的 IP 地址。

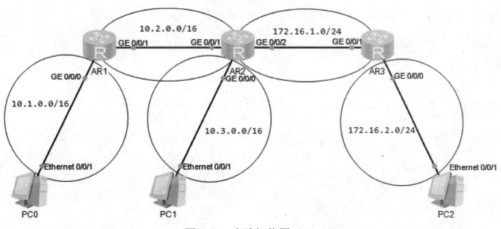

图 4-14 实验拓扑图（1）

表 4-5 各设备的 IP 地址配置（1）

设备名称	接口	IP 地址	默认网关
AR1	GE0/0/0	10.1.0.254/16	
	GE0/0/1	10.2.0.1/16	
AR2	GE0/0/0	10.3.0.254/16	
	GE0/0/1	10.2.0.2/16	
	GE0/0/2	172.16.1.1/24	
AR3	GE0/0/0	172.16.2.254/24	
	GE0/0/1	172.16.1.2/24	
PC0	Ethernet0/0/1	10.1.0.1/16	10.1.0.254/16
PC1	Ethernet0/0/1	10.3.0.1/16	10.3.0.254/16
PC2	Ethernet0/0/1	172.16.2.1/24	172.16.2.254/24

（2）在路由器上配置 RIPv1 路由。

在 AR1 上配置 RIPv1 路由。配置如下：

```
[AR1]rip 1
[AR1-rip-1]version 1
[AR1-rip-1]network 10.0.0.0
```

在 AR2 上配置 RIPv1 路由。配置如下：

```
[AR2]rip 1
[AR2-rip-1]version 1
[AR2-rip-1]network 10.0.0.0
[AR2-rip-1]network 172.16.0.0
```

在 AR3 上配置 RIPv1 路由。配置如下：

```
[AR3]rip 1
[AR3-rip-1]version 1
[AR3-rip-1]network 172.16.0.0
```

（3）查看路由器的路由表。

查看 AR1 的路由表，结果如下：

```
[AR1]display ip routing-table
Route Flags: R - relay, D - download to fib
------------------------------------------------------------------------
Routing Tables: Public
        Destinations : 12       Routes : 12
```

Destination/Mask	Proto	Pre	Cost	Flags	NextHop	Interface
10.1.0.0/16	Direct	0	0	D	10.1.0.254	GigabitEthernet0/0/0
10.1.0.254/32	Direct	0	0	D	127.0.0.1	GigabitEthernet0/0/0
0.1.255.255/32	Direct	0	0	D	127.0.0.1	GigabitEthernet0/0/0
10.2.0.0/16	Direct	0	0	D	10.2.0.1	GigabitEthernet0/0/1
10.2.0.1/32	Direct	0	0	D	127.0.0.1	GigabitEthernet0/0/1
10.2.255.255/32	Direct	0	0	D	127.0.0.1	GigabitEthernet0/0/1
10.3.0.0/16	RIP	100	1	D	10.2.0.2	GigabitEthernet0/0/1
127.0.0.0/8	Direct	0	0	D	127.0.0.1	InLoopBack0
127.0.0.1/32	Direct	0	0	D	127.0.0.1	InLoopBack0
127.255.255.255/32	Direct	0	0	D	127.0.0.1	InLoopBack0
172.16.0.0/16	RIP	100	1	D	10.2.0.2	GigabitEthernet0/0/1
255.255.255.255/32	Direct	0	0	D	127.0.0.1	InLoopBack0

该拓扑中共有五个网络，AR1 中只显示了四个，其中 172.16.0.0/16 是 AR2 将 172.16.1.0/24 和 172.16.2.0/24 汇总的结果，汇总后再发送给 AR1。路由汇总默认是开启的，也可以使用命令 undo summary 将自动汇总关闭。

查看路由器 AR1 的 RIP 配置信息及 RIP 的一些参数，结果如下：

```
[AR1]dis rip 1
Public VPN-instance
    RIP process      : 1
    RIP version      : 1
    Preference       : 100
    Checkzero        : Enabled
    Default-cost     : 0
    Summary          : Disabled
    Host-route       : Enabled
    Maximum number of balanced paths : 8
 Update time   : 30 sec     Age time : 180 sec      //RIP 的时间参数。
    Garbage-collect time : 120 sec
    Graceful restart     : Disabled
    BFD                  : Disabled
    Silent-interfaces    : None
    Default-route        : Disabled
    Verify-source        : Enabled
    Networks             : 10.0.0.0
    //路由的网络号。
    Configured peers             : None
```

```
            Number of routes in database  : 5
            Number of interfaces enabled  : 2
            Triggered updates sent        : 1
            Number of route changes       : 2
            Number of replies to queries  : 1
       Number of routes in ADV DB         : 4
```

查看 AR2 的路由表，结果如下：

```
[AR2]display ip routing-table
Route Flags: R - relay, D - download to fib
------------------------------------------------------------------------
Routing Tables: Public
        Destinations : 15      Routes : 15

Destination/Mask      Proto   Pre  Cost  Flags  NextHop       Interface
       10.1.0.0/16    RIP     100  1       D    10.2.0.1      GigabitEthernet0/0/1
       10.2.0.0/16    Direct  0    0       D    10.2.0.2      GigabitEthernet0/0/1
       10.2.0.2/32    Direct  0    0       D    127.0.0.1     GigabitEthernet0/0/1
  10.2.255.255/32     Direct  0    0       D    127.0.0.1     GigabitEthernet0/0/1
       10.3.0.0/16    Direct  0    0       D    10.3.0.254    GigabitEthernet0/0/0
     10.3.0.254/32    Direct  0    0       D    127.0.0.1     GigabitEthernet0/0/0
  10.3.255.255/32     Direct  0    0       D    127.0.0.1     GigabitEthernet0/0/0
      127.0.0.0/8     Direct  0    0       D    127.0.0.1     InLoopBack0
      127.0.0.1/32    Direct  0    0       D    127.0.0.1     InLoopBack0
 127.255.255.255/32   Direct  0    0       D    127.0.0.1     InLoopBack0
     172.16.1.0/24    Direct  0    0       D    172.16.1.1    GigabitEthernet0/0/2
     172.16.1.1/32    Direct  0    0       D    127.0.0.1     GigabitEthernet0/0/2
   172.16.1.255/32    Direct  0    0       D    127.0.0.1     GigabitEthernet0/0/2
     172.16.2.0/24    RIP     100  1       D    172.16.1.2    GigabitEthernet0/0/2
 255.255.255.255/32   Direct  0    0       D    127.0.0.1     InLoopBack0
```

通过查看路由表可知，AR2 中有五条路由，其中网络 10.1.0.0/16 和 172.16.2.0/24 是通过 RIP 学习到的。

查看 AR3 的路由表，结果如下：

```
[AR3]display ip routing-table
Route Flags: R - relay, D - download to fib
------------------------------------------------------------------------
Routing Tables: Public
        Destinations : 11      Routes : 11
```

Destination/Mask	Proto	Pre	Cost	Flags	NextHop	Interface
10.0.0.0/8	RIP	100	1	D	172.16.1.1	GigabitEthernet0/0/1
127.0.0.0/8	Direct	0	0	D	127.0.0.1	InLoopBack0
127.0.0.1/32	Direct	0	0	D	127.0.0.1	InLoopBack0
127.255.255.255/32	Direct	0	0	D	127.0.0.1	InLoopBack0
172.16.1.0/24	Direct	0	0	D	172.16.1.2	GigabitEthernet0/0/1
172.16.1.2/32	Direct	0	0	D	127.0.0.1	GigabitEthernet0/0/1
172.16.1.255/32	Direct	0	0	D	127.0.0.1	GigabitEthernet0/0/1
172.16.2.0/24	Direct	0	0	D	172.16.2.254	GigabitEthernet0/0/0
172.16.2.254/32	Direct	0	0	D	127.0.0.1	GigabitEthernet0/0/0
172.16.2.255/32	Direct	0	0	D	127.0.0.1	GigabitEthernet0/0/0
255.255.255.255/32	Direct	0	0	D	127.0.0.1	InLoopBack0

AR3 中只有一条路由是通过 RIP 学习到的，而且学到的是汇总后的路由。

（4）查看 RIP 路由的动态更新。查看 AR1 的 RIP 路由的动态更新，结果如下：

```
<AR1>terminal monitor
<AR1>terminal debugging
<AR1>debugging rip 1
Periodic timer expired for interface GigabitEthernet0/0/1
//从 GE0/0/1 接口收到来自 10.2.0.2（路由器 AR2）的 RIPv1 的更新分组，内容如下。
Job Periodic Update is created
Periodic timer expired for interface GigabitEthernet0/0/1 (10.2.0.1) and its added to periodic update queue
Job Periodic Update is scheduled for interface GigabitEthernet0/0/1
Periodic UpdateCompleted for interface GigabitEthernet0/0/1, Time = 0 Ms
Interface GigabitEthernet0/0/1 (10.2.0.1) is deleted from the periodic update queue
Sending v1 response on GigabitEthernet0/0/1 from 10.2.0.1 with 1 RTE
Sending response on interface GigabitEthernet0/0/1 from 10.2.0.1 to 255.255.255.255Version 1, Cmdresponse, Length 24
Dest 10.1.0.0, Cost 1
Periodic timer expired for interface GigabitEthernet0/0/0
Job Periodic Update is created
Periodic timer expired for interface GigabitEthernet0/0/0 (10.1.0.254) and its added to periodic update queue
Job Periodic Update is scheduled for interface GigabitEthernet0/0/0
Periodic UpdateCompleted for interface GigabitEthernet0/0/0, Time = 0 Ms
Interface GigabitEthernet0/0/0 (10.1.0.254) is deleted from the periodic
```

```
update queue
    Sending v1 response on GigabitEthernet0/0/0 from 10.1.0.254 with 4 RTEs
    Sending response on interface GigabitEthernet0/0/0 from 10.1.0.254 to
255.255.255.255
    //路由器通过 GE0/0/0 接口广播发送生成的 RIPv1 的更新分组，内容如上所示。请注意观察，这个
    //更新分组是路由器刚刚收到 AR1 传来的更新分组后，根据距离向量算法重新生成的路由[详细算法
    //请参见教材《计算机网络（第 8 版）》4.6.2 节]，然后将这个更新分组转发给邻居。这个邻居在
    //这里是 PC0，实际上，对于主机来说，并不需要接收这样的路由更新。所以，可以将 GE0/0/0 接
    //口设置为静默接口，这样，路由器就不会从此接口发送路由更新了，但是依旧可以接收更新分组。
    //比如，运行 silent-interface GigabitEthernet 0/0/0 命令后，再次查看 RIP 的动态更新，
    //将不会有从 GE0/0/0 接口发送的更新分组，请自行验证。
    Packet: Version 1, Cmdresponse, Length 84
    Dest 10.1.0.0, Cost 1
    Dest 10.2.0.0, Cost 1
    Dest 10.3.0.0, Cost 2
    Dest 172.16.0.0, Cost 2
    Receiving v1 response on GigabitEthernet0/0/1 from 10.2.0.2 with 2 RTEs
    Receive responsefrom 10.2.0.2 on GigabitEthernet0/0/1
```

另外，路由更新信息会霸屏，不需要时应及时将其关闭，运行如下命令即可：

```
<AR1>quit
```

请读者自行解释并验证路由器 AR1 和 AR2 的路由更新信息。

（5）由 PC0 去 ping PC1 和 PC2，可以 ping 通，请读者自行验证。

5. RIPv2 实验步骤

为了体现与 RIPv1 的区别，这里的网络采用变长子网掩码来设计。拓扑中包含五个网络，其网络划分如表 4-6 所示。

表 4-6 网络划分

网络地址	子网掩码	第一个 IP 地址	最后一个 IP 地址
192.168.1.32	255.255.255.224	192.168.1.33	192.168.1.62
192.168.1.64	255.255.255.224	192.168.1.65	192.168.1.94
192.168.1.96	255.255.255.252	192.168.1.97	192.168.1.98
192.168.1.128	255.255.255.224	192.168.1.129	192.168.1.158
192.168.1.160	255.255.255.224	192.168.1.161	192.168.1.190

（1）按图 4-15 所示布置拓扑，并按表 4-7 配置各设备的 IP 地址。这里将网络的第一个可用的 IP 地址设为 PC 的 IP 地址，将网络的最后一个可用 IP 地址设为网关，连接两台路由器的网络为/30 的地址。

第 4 章　网络层

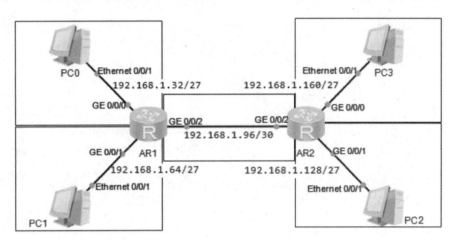

图 4-15　实验拓扑图（2）

表 4-7　各设备的 IP 地址配置（2）

设备名称	接口	IP 地址	默认网关
AR1	GE0/0/0	192.168.1.62/27	
	GE0/0/1	192.168.1.94/27	
	GE0/0/2	192.168.1.97/30	
AR2	GE0/0/0	192.168.1.190/27	
	GE0/0/1	192.168.1.158/27	
	GE0/0/2	192.168.1.98/30	
PC0	Ethernet0/0/1	192.168.1.33/27	192.168.1.62/27
PC1	Ethernet0/0/1	192.168.1.65/27	192.168.1.94/27
PC2	Ethernet0/0/1	192.168.1.129/27	192.168.1.158/27
PC3	Ethernet0/0/1	192.168.1.161/27	192.168.1.190/27

AR1 的 IP 地址配置如下：

```
<AR1>sy
Enter system view, return user view with Ctrl+Z.
[AR1]int g0/0/0
[AR1-GigabitEthernet0/0/0]ip add 192.168.1.62 27
[AR1-GigabitEthernet0/0/0]q
[AR1]int g0/0/1
[AR1-GigabitEthernet0/0/1]ip add 192.168.1.94 27
[AR1-GigabitEthernet0/0/1]q
[AR1]int g0/0/2
[AR1-GigabitEthernet0/0/2]ip add 192.168.1.97 30
```

AR2 的 IP 地址配置如下：

```
<AR2>sy
```

```
Enter system view, return user view with Ctrl+Z.
[AR2]int g0/0/0
[AR2-GigabitEthernet0/0/0]ip add 192.168.1.190 27
[AR2-GigabitEthernet0/0/0]q
[AR2]int g0/0/1
[AR2-GigabitEthernet0/0/1]ip add 192.168.1.158 27
[AR2-GigabitEthernet0/0/1]q
[AR2]int g0/0/2
[AR2-GigabitEthernet0/0/2]ip add 192.168.1.98 30
```

（2）在路由器上配置 RIPv2 路由。

在 AR1 上配置 RIPv2 路由。配置如下：

```
[AR1]rip 1
[AR1-rip-1]version 2
[AR1-rip-1]network 192.168.1.0
[AR1-rip-1]undo summary
[AR1-rip-1]silent-interface GigabitEthernet 0/0/0
[AR1-rip-1]silent-interface GigabitEthernet 0/0/1
```

在 AR2 上配置 RIPv2 路由。配置如下：

```
[AR2]rip 1
[AR2-rip-1]version 2
[AR2-rip-1]network 192.168.1.0
[AR2-rip-1]undo summary
[AR2-rip-1]silent-interface GigabitEthernet 0/0/0
[AR2-rip-1]silent-interface GigabitEthernet 0/0/1
```

这里将两台路由器 RIP 的自动汇总关闭（默认是开启的），并设置连接主机的两个接口为静默接口，不向主机发送 RIP 路由更新分组。

（3）查看路由器的路由表。

查看 AR1 的路由表，结果如下：

```
[AR1]dis ip routing-table
Route Flags: R - relay, D - download to fib
------------------------------------------------------------------
Routing Tables: Public
         Destinations : 15       Routes : 15
  Destination/Mask    Proto  Pre  Cost  Flags NextHop         Interface
       127.0.0.0/8    Direct  0    0     D    127.0.0.1       InLoopBack0
```

127.0.0.1/32	Direct	0	0	D	127.0.0.1	InLoopBack0
127.255.255.255/32	Direct	0	0	D	127.0.0.1	InLoopBack0
192.168.1.32/27	Direct	0	0	D	192.168.1.62	GigabitEthernet0/0/0
192.168.1.62/32	Direct	0	0	D	127.0.0.1	GigabitEthernet0/0/0
192.168.1.63/32	Direct	0	0	D	127.0.0.1	GigabitEthernet0/0/0
192.168.1.64/27	Direct	0	0	D	192.168.1.94	GigabitEthernet0/0/1
192.168.1.94/32	Direct	0	0	D	127.0.0.1	GigabitEthernet0/0/1
192.168.1.95/32	Direct	0	0	D	127.0.0.1	GigabitEthernet0/0/1
192.168.1.96/30	Direct	0	0	D	192.168.1.97	GigabitEthernet0/0/2
192.168.1.97/32	Direct	0	0	D	127.0.0.1	GigabitEthernet0/0/2
192.168.1.99/32	Direct	0	0	D	127.0.0.1	GigabitEthernet0/0/2
192.168.1.128/27	RIP	100	1	D	192.168.1.98	GigabitEthernet0/0/2
192.168.1.160/27	RIP	100	1	D	192.168.1.98	GigabitEthernet0/0/2
255.255.255.255/32	Direct	0	0	D	127.0.0.1	InLoopBack0

查看 AR2 的路由表，结果如下：

```
[AR2]dis ip routing-table
Route Flags: R - relay, D - download to fib
------------------------------------------------------------------
Routing Tables: Public
         Destinations : 15        Routes : 15
Destination/Mask    Proto  Pre  Cost  Flags  NextHop         Interface
```

Destination/Mask	Proto	Pre	Cost	Flags	NextHop	Interface
127.0.0.0/8	Direct	0	0	D	127.0.0.1	InLoopBack0
127.0.0.1/32	Direct	0	0	D	127.0.0.1	InLoopBack0
127.255.255.255/32	Direct	0	0	D	127.0.0.1	InLoopBack0
192.168.1.32/27	RIP	100	1	D	192.168.1.97	GigabitEthernet0/0/2
192.168.1.64/27	RIP	100	1	D	192.168.1.97	GigabitEthernet0/0/2
192.168.1.96/30	Direct	0	0	D	192.168.1.98	GigabitEthernet0/0/2
192.168.1.98/32	Direct	0	0	D	127.0.0.1	GigabitEthernet0/0/2
192.168.1.99/32	Direct	0	0	D	127.0.0.1	GigabitEthernet0/0/2
192.168.1.128/27	Direct	0	0	D	192.168.1.158	GigabitEthernet0/0/1
192.168.1.158/32	Direct	0	0	D	127.0.0.1	GigabitEthernet0/0/1
192.168.1.159/32	Direct	0	0	D	127.0.0.1	GigabitEthernet0/0/1
192.168.1.160/27	Direct	0	0	D	192.168.1.190	GigabitEthernet0/0/0
192.168.1.190/32	Direct	0	0	D	127.0.0.1	GigabitEthernet0/0/0
192.168.1.191/32	Direct	0	0	D	127.0.0.1	GigabitEthernet0/0/0
255.255.255.255/32	Direct	0	0	D	127.0.0.1	InLoopBack0

通过查看路由表可知，两台路由器的路由表均包含五个网络，其中三个为直连网络，另

外两个是通过 RIP 得到的。

（4）查看 RIP 路由的动态更新。

查看 AR1 的 RIP 路由的动态更新，结果如下：

```
<AR1>terminal monitor
<AR1>terminal debugging
<AR1>debugging rip 1
Periodic timer expired for interface GigabitEthernet0/0/1
Periodic timer expired for interface GigabitEthernet0/0/0
Receiving v2 response on GigabitEthernet0/0/2 from 192.168.1.98 with 2 RTEs
Receive response from 192.168.1.98 on GigabitEthernet0/0/2
Version 2, Cmdresponse, Length 44
Dest 192.168.1.128/27,Nexthop 0.0.0.0, Cost 1, Tag 0
Dest 192.168.1.160/27,Nexthop 0.0.0.0, Cost 1, Tag 0
//从 GE0/0/2 接口收到 RIPv2 的更新分组，注意是/27 的网络。

Periodic timer expired for interface GigabitEthernet0/0/2
Job Periodic Upate is created
Periodic timer expired for interface GigabitEthernet0/0/2 (192.168.1.97) and its added to periodic update queue
Job Periodic Update is scheduled for interface GigabitEthernet0/0/2
Periodic UpdateCompleted for interface GigabitEthernet0/0/2, Time = 0 Ms
Interface GigabitEthernet0/0/2 (192.168.1.97) is deleted from the periodic update queue
Sending v2 response on GigabitEthernet0/0/2 from 192.168.1.97 with 2 RTEs
Sending response on interface GigabitEthernet0/0/2 from 192.168.1.97 to 224.0.0.9
Version 2, Cmdresponse, Length 44
Dest 192.168.1.32/27, Nexthop 0.0.0.0, Cost 1, Tag 0
Dest 192.168.1.64/27, Nexthop 0.0.0.0, Cost 1, Tag 0
//RIPv2 使用组播发送自己的路由信息分组，而 RIPv1 是通过广播发送的。注意更新分组，RIP
//使用了水平切割。另外，因为将连接主机的接口定义为静默接口，所以此处没有发往主机的路由
//更新分组。
```

查看 AR2 的 RIP 路由的动态更新，结果如下：

```
<AR2>terminal monitor
<AR2>terminal debugging
<AR2>debugging rip 1
Periodic timer expired for interface GigabitEthernet0/0/0
```

```
    Periodic timer expired for interface GigabitEthernet0/0/1
    Periodic timer expired for interface GigabitEthernet0/0/2
    Job Periodic Update is created
    Periodic timer expired for interface GigabitEthernet0/0/2 (192.168.1.98) and
its added to periodic update queue
    Job Periodic Update is scheduled for interface GigabitEthernet0/0/2
    Periodic UpdateCompleted for interface GigabitEthernet0/0/2, Time = 0 Ms
    Interface GigabitEthernet0/0/2 (192.168.1.98) is deleted from the periodic
update queue
    Sending v2 response on GigabitEthernet0/0/2 from 192.168.1.98 with 2 RTEs
    Sending responseon interface GigabitEthernet0/0/2 from 192.168.1.98 to
224.0.0.9
    Version 2, Cmd response, Length 44
    Dest 192.168.1.128/27, Nexthop 0.0.0.0, Cost 1, Tag 0
    Dest 192.168.1.160/27, Nexthop 0.0.0.0, Cost 1, Tag 0
    Receiving v2 response on GigabitEthernet0/0/2 from 192.168.1.97 with 2 RTEs
    Receive response from 192.168.1.97 on GigabitEthernet0/0/2
    Version 2, Cmdresponse, Length 44
    Dest 192.168.1.32/27, Nexthop 0.0.0.0, Cost 1, Tag 0
    Dest 192.168.1.64/27, Nexthop 0.0.0.0, Cost 1, Tag 0
```

请读者自行分析 AR1 中的路由动态更新，并结合距离向量算法验证路由器的路由表。

（5）此时，主机间两两都可以 ping 通，请读者自行验证。

本例中，RIP 配置只需要宣告 192.168.1.0/24 的网段（主类网段）即可，通过查看路由表可知，通过 RIP 学习到的是 VLSM 网段。

实验五：内部网关协议 OSPF 实验

1. 实验目的

（1）理解 OSPF。
（2）掌握 OSPF 的配置方法。
（3）掌握查看 OSPF 相关信息的方法。

2. 基础知识

OSPF（Open Shortest Path First）的中文名称为开放最短路径优先，是一个内部网关协议。OSPF 用于在单一自治系统（AS）中决策路由，使用迪杰斯特拉（Dijkstra）提出的最短路径算法（SPF）生成路由表。OSPF 分为 OSPFv2 和 OSPFv3 两个版本，其中 OSPFv2 用于 IPv4 网络，OSPFv3 用于 IPv6 网络。

OSPF 有以下特点。

（1）OSPF 是链路状态协议。OSPF 将连接两台路由器的链路状态归结为一个度量或代价，以此来表示链路的时延、带宽、费用、距离等，可以由管理员配置其大小，范围为 1~65 535，但很多时候并没有显式的配置，这时代价被默认为 100Mb/s 带宽。而 RIP 中只用跳数来衡量距离，不考虑链路的状态。

（2）OSPF 使用最短路径算法计算路由。OSPF 中的每一台路由器都会维护一个本区域的拓扑结构图（LSDB），路由器依据拓扑结构图中的节点和链路代价计算出一棵以自己为根的最短路径树，根据这棵树，就可以找到去往各目的网络的最短路径，并生成自己的路由表。由于 OSPF 使用最短路径树算法计算路由，因此从算法本身保证了不会生成自环路由。

（3）OSPF 使用层次结构的区域划分。

① 划分区域的理由。由于每一台路由器都会维护一个全网的拓扑结构，因此当网络规模达到一定程度时，拓扑结构势必形成一个庞大的数据库，这样会给 OSPF 的计算最短路径算法带来比较大的压力。为了能够降低计算的复杂程度，OSPF 采用分区域计算，即将网络中所有 OSPF 路由器划分成不同的区域，每个区域负责各自区域精确的链路状态通告（LSA）传递与路由计算，然后将一个区域的 LSA 简化和汇总之后转发到另外一个区域，这样一来，在区域内部，拥有网络精确的 LSA，而在不同区域，则传递简化的 LSA。

② 区域的层次。区域分为骨干区域（BackBone Area）和标准区域（Normal Area），其中，骨干区域位于顶层。一般来说，所有的标准区域应该直接和骨干区域相连，标准区域间的通信分组要通过骨干区域路由转发，标准区域只能和骨干区域交换 LSA，标准区域相互之间即使直连也无法交换 LSA。

③ 区域的命名可以采用整数数字，如 1、2、3 等；也可以采用 32 位 IP 地址的形式，如 0.0.0.1、0.0.0.2。需要注意的是，骨干区域只能被命名为 0 区域，而不能是其他区域。

④ OSPF 区域中路由器的区别。OSPF 区域是基于路由器的接口划分的，而不是基于整台路由器划分的，这样，一台路由器既可以属于单个区域，也可以属于多个区域。如果一台 OSPF 路由器属于单个区域，即该路由器所有接口都属于同一个区域，那么这台路由器称为区域内部路由器（IR）。如果一台 OSPF 路由器属于多个区域，即该路由器的接口不都属于同一个区域，那么这台路由器称为区域边界路由器（ABR），ABR 可以将一个区域的 LSA 汇总后转发至另一个区域。如果一台 OSPF 路由器代表本自治系统连接到外部自治系统的路由器上，那么这台路由器称为自治系统边界路由器（ASBR），由 ASBR 代表本 AS 与外部进行路由。

（4）负载平衡。OSPF 支持到同一目的地址的多条等值路由，以此实现负载平衡。

（5）收敛速度快。如果网络的拓扑结构发生变化，那么 OSPF 会立即泛洪发送更新报文，让这一变化快速在自治系统中同步，使路由快速收敛。

（6）OSPF 在描述路由时携带网段的掩码信息，所以其支持 VLSM 和 CIDR（无分类编址）。

（7）支持验证。OSPF 支持基于接口的报文验证以保证路由计算的安全性。

（8）OSPF 自身的开销控制较小。

① 使用定期发送 hello 报文的方式来发现和维护邻居关系。hello 报文非常小，减少了网

络资源的消耗。

② 在可以多址访问的网络中，如广播网络（以太网）或 NBMA 网络（如 x.25、帧中继），通过选举 DR 和 BDR，使同网段的路由器之间的路由交换次数减少。DR 从邻居那里收到更新后，通过 LSA 通告给局域网上的所有邻居，使同一个局域网的拓扑结构图都相同。

③ 在广播网络中，使用组播地址发送报文，这样可以减少对其他不运行 OSPF 的网络设备的干扰。

④ 在 ABR 上支持路由聚合，进一步减少区域间的路由信息传递。

关于 OSPF 更详细的内容请参见教材《计算机网络（第 8 版）》4.6.3 节。

本实验配置命令如表 4-8 所示。

表 4-8 本实验配置命令

命令格式	命令说明
ospf 进程号 router-id A.B.C.D	系统模式下进入 OSPF 配置模式。配置路由器的 router-id（以 IP 地址的形式）
area 区域号	进入区域
network 网络地址 通配符	通告直连网络，通配符为网络掩码的反码
display ospf peer	查看路由协议配置与统计信息
display ospf brief	查看 OSPF 进程及区域细节的数据
display ospf lsdb	查看路由器 OSPF 数据库信息
display ospf interface	查看接口 OSPF 信息
display ospf peer brief	查看 OSPF 邻居信息
debugging ospf packet	查看 OSPF 报文

3. 实验流程

实验流程如图 4-16 所示。

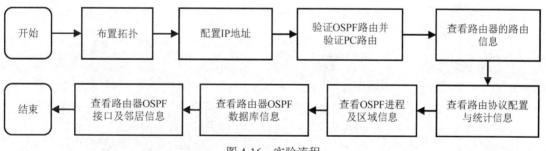

图 4-16 实验流程

4. 单区域 OSPF 路由配置实验步骤

单区域 OSPF 应用于网络规模不大，只使用 area 0 一个区域就可满足需求的情况。

（1）布置拓扑。如图 4-17 所示，四台路由器连接了五个网段，全部属于 area 0，其 IP 地址配置如表 4-9 所示。

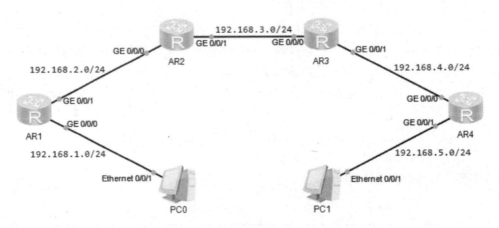

图 4-17 单区域 OSPF 拓扑图

表 4-9 各设备的 IP 地址配置（1）

设备名称	接口	IP 地址	默认网关	区域（area）
AR1	GE0/0/0	192.168.1.254/24		0
	GE0/0/1	192.168.2.1/24		0
AR2	GE0/0/0	192.168.2.2/24		0
	GE0/0/1	192.168.3.2/24		0
AR3	GE0/0/0	192.168.3.3/24		0
	GE0/0/1	192.168.4.3/24		0
AR4	GE0/0/0	192.168.4.4/24		0
	GE0/0/1	192.168.5.254/24		0
PC0	Ethernet0/0/1	192.168.1.1/24	192.168.1.254	
PC1	Ethernet0/0/1	192.168.5.1/24	192.168.5.254	

（2）配置路由器的 OSPF 路由，将全部路由器接口配置到 area 0。

AR1 中 OSPF 路由配置如下：

```
[AR1]ospf 1 router-id 1.1.1.1
//进入 OSPF 路由配置模式，进程号为 1。进程号可以设置为 1~65 535 中的任意值，但其只具有本
//地意义，不需要在路由器之间匹配，也就是说，进程号可以不同。进程号用来区分本路由器上运行
//的不同 OSPF 进程，如本路由器可能是两个自治系统的边界路由器。
//设置路由器 ID 为 1.1.1.1，也就是说在 OSPF 中给路由器起个名字，用来标识该路由器，这是必
//要的。其实，即便不显式地设置路由器 ID，OSPF 也会按如下顺序认可一个地址作为路由器 ID：
//首先是路由器最大的环回接口的 IP 地址，其次是最大的活动物理接口的 IP 地址。当然，最好是使
//用命令指定路由器 ID。
[AR1-ospf-1]area 0
//说明该网络属于哪个区域，这一点很重要，因为 OSPF 的 LSA 和区域有直接关系。
[AR1-ospf-1-area-0.0.0.0]network 192.168.1.0 0.0.0.255
```

```
[AR1-ospf-1-area-0.0.0.0]network 192.168.2.0 0.0.0.255
//宣告本路由器的直连网络，命令中使用了通配符，通配符为网络掩码的反码，所以，OSPF支持VLSM
//和CIDR。
```

AR2 中 OSPF 路由配置如下：

```
[AR2]ospf 1 router-id 2.2.2.2
[AR2-ospf-1]area 0
[AR2-ospf-1-area-0.0.0.0]network 192.168.2.0 0.0.0.255
[AR2-ospf-1-area-0.0.0.0]network 192.168.3.0 0.0.0.255
```

AR3 中 OSPF 路由配置如下：

```
[AR3]ospf 1 router-id 3.3.3.3
[AR3-ospf-1]area 0
[AR3-ospf-1-area-0.0.0.0]network 192.168.3.0 0.0.0.255
[AR3-ospf-1-area-0.0.0.0]network 192.168.4.0 0.0.0.255
```

AR4 中 OSPF 路由配置如下：

```
[AR4]ospf 1 router-id 5.5.5.5
[AR4-ospf-1]area 0
[AR4-ospf-1-area-0.0.0.0]network 192.168.4.0 0.0.0.255
[AR4-ospf-1-area-0.0.0.0]network 192.168.5.0 0.0.0.255
```

（3）验证路由可达。由 PC0 ping PC1，如图 4-18 所示，结果是通的，说明 OSPF 路由配置正确。

（4）查看路由器的路由表。

查看 AR1 的路由表，结果如下：

```
[AR1]display ip routing-table
Route Flags: R - relay, D - download to fib
------------------------------------------------------------------
Routing Tables: Public
         Destinations : 13       Routes : 13
   Destination/Mask    Proto   Pre  Cost  Flags  NextHop         Interface
        127.0.0.0/8    Direct   0    0      D    127.0.0.1       InLoopBack0
        127.0.0.1/32   Direct   0    0      D    127.0.0.1       InLoopBack0
  127.255.255.255/32   Direct   0    0      D    127.0.0.1       InLoopBack0
      192.168.1.0/24   Direct   0    0      D    192.168.1.254   GigabitEthernet0/0/0
    192.168.1.254/32   Direct   0    0      D    127.0.0.1       GigabitEthernet0/0/0
    192.168.1.255/32   Direct   0    0      D    127.0.0.1       GigabitEthernet0/0/0
```

```
     192.168.2.0/24    Direct  0   0    D   192.168.2.1   GigabitEthernet0/0/1
     192.168.2.1/32    Direct  0   0    D   127.0.0.1     GigabitEthernet0/0/1
     192.168.2.255/32  Direct  0   0    D   127.0.0.1     GigabitEthernet0/0/1
     192.168.3.0/24    OSPF    10  2    D   192.168.2.2   GigabitEthernet0/0/1
     192.168.4.0/24    OSPF    10  3    D   192.168.2.2   GigabitEthernet0/0/1
     192.168.5.0/24    OSPF    10  4    D   192.168.2.2   GigabitEthernet0/0/1
 255.255.255.255/32    Direct  0   0    D   127.0.0.1     InLoopBack0
```

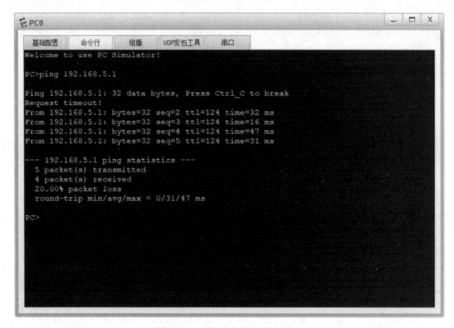

图 4-18 验证路由可达（1）

通过查看路由表可知，AR1 中有五条路由，对应拓扑中的五个网段，其中 192.168.1.0/24 和 192.168.2.0/24 为路由器的直连路由，其他三个 Proto 为 OSPF 的网段为通过 OSPF 学到的路由。

下面为其他路由器的路由表，请读者自行分析。

查看 AR2 的路由表，结果如下：

```
[AR2]display ip routing-table
Route Flags: R - relay, D - download to fib
-----------------------------------------------------------------------
Routing Tables: Public
         Destinations : 13        Routes : 13
   Destination/Mask    Proto  Pre  Cost  Flags  NextHop       Interface
        127.0.0.0/8    Direct  0   0      D     127.0.0.1     InLoopBack0
        127.0.0.1/32   Direct  0   0      D     127.0.0.1     InLoopBack0
    127.255.255.255/32 Direct  0   0      D     127.0.0.1     InLoopBack0
```

```
     192.168.1.0/24    OSPF    10   2    D    192.168.2.1    GigabitEthernet0/0/0
     192.168.2.0/24    Direct  0    0    D    192.168.2.2    GigabitEthernet0/0/0
     192.168.2.2/32    Direct  0    0    D    127.0.0.1      GigabitEthernet0/0/0
   192.168.2.255/32    Direct  0    0    D    127.0.0.1      GigabitEthernet0/0/0
     192.168.3.0/24    Direct  0    0    D    192.168.3.2    GigabitEthernet0/0/1
     192.168.3.2/32    Direct  0    0    D    127.0.0.1      GigabitEthernet0/0/1
   192.168.3.255/32    Direct  0    0    D    127.0.0.1      GigabitEthernet0/0/1
     192.168.4.0/24    OSPF    10   2    D    192.168.3.3    GigabitEthernet0/0/1
     192.168.5.0/24    OSPF    10   3    D    192.168.3.3    GigabitEthernet0/0/1
   255.255.255.255/32  Direct  0    0    D    127.0.0.1      InLoopBack0
```

查看 AR3 的路由表，结果如下：

```
[AR3]display ip routing-table
Route Flags: R - relay, D - download to fib
------------------------------------------------------------------
Routing Tables: Public
        Destinations : 13       Routes : 13
Destination/Mask      Proto   Pre  Cost Flags NextHop        Interface
      127.0.0.0/8     Direct  0    0    D    127.0.0.1      InLoopBack0
      127.0.0.1/32    Direct  0    0    D    127.0.0.1      InLoopBack0
  127.255.255.255/32  Direct  0    0    D    127.0.0.1      InLoopBack0
     192.168.1.0/24   OSPF    10   3    D    192.168.3.2    GigabitEthernet0/0/0
     192.168.2.0/24   OSPF    10   2    D    192.168.3.2    GigabitEthernet0/0/0
     192.168.3.0/24   Direct  0    0    D    192.168.3.3    GigabitEthernet0/0/0
     192.168.3.3/32   Direct  0    0    D    127.0.0.1      GigabitEthernet0/0/0
   192.168.3.255/32   Direct  0    0    D    127.0.0.1      GigabitEthernet0/0/0
     192.168.4.0/24   Direct  0    0    D    192.168.4.3    GigabitEthernet0/0/1
     192.168.4.3/32   Direct  0    0    D    127.0.0.1      GigabitEthernet0/0/1
   192.168.4.255/32   Direct  0    0    D    127.0.0.1      GigabitEthernet0/0/1
     192.168.5.0/24   OSPF    10   2    D    192.168.4.4    GigabitEthernet0/0/1
  255.255.255.255/32  Direct  0    0    D    127.0.0.1      InLoopBack0
```

查看 AR4 的路由表，结果如下：

```
[AR4]display ip routing-table
Route Flags: R - relay, D - download to fib
------------------------------------------------------------------
Routing Tables: Public
        Destinations : 13       Routes : 13
```

```
Destination/Mask         Proto   Pre  Cost  Flags  NextHop        Interface
        127.0.0.0/8      Direct  0    0     D      127.0.0.1      InLoopBack0
        127.0.0.1/32     Direct  0    0     D      127.0.0.1      InLoopBack0
   127.255.255.255/32    Direct  0    0     D      127.0.0.1      InLoopBack0
      192.168.1.0/24     OSPF    10   4     D      192.168.4.3    GigabitEthernet0/0/0
      192.168.2.0/24     OSPF    10   3     D      192.168.4.3    GigabitEthernet0/0/0
      192.168.3.0/24     OSPF    10   2     D      192.168.4.3    GigabitEthernet0/0/0
      192.168.4.0/24     Direct  0    0     D      192.168.4.4    GigabitEthernet0/0/0
      192.168.4.4/32     Direct  0    0     D      127.0.0.1      GigabitEthernet0/0/0
    192.168.4.255/32     Direct  0    0     D      127.0.0.1      GigabitEthernet0/0/0
      192.168.5.0/24     Direct  0    0     D      192.168.5.254  GigabitEthernet0/0/1
    192.168.5.254/32     Direct  0    0     D      127.0.0.1      GigabitEthernet0/0/1
    192.168.5.255/32     Direct  0    0     D      127.0.0.1      GigabitEthernet0/0/1
   255.255.255.255/32    Direct  0    0     D      127.0.0.1      InLoopBack0
```

（5）查看OSPF中各区域邻居的信息，以AR1为例，结果如下：

```
[AR1]display ospf peer
            OSPF Process 1 with Router ID 1.1.1.1
                      Neighbors
 Area 0.0.0.0 interface 192.168.2.1(GigabitEthernet0/0/1)'s neighbors
 Router ID: 2.2.2.2        Address: 192.168.2.2
   State: Full  Mode:Nbr is Master  Priority: 1
   DR: 192.168.2.1 BDR: 192.168.2.2 MTU: 0
   Dead timer due in 30 sec
   Retrans timer interval: 5
   Neighbor is up for 00:31:16
 Authentication Sequence: [ 0 ]
```

（6）查看OSPF进程及区域细节的数据，以AR1为例，结果如下：

```
[AR1]display ospf brief
            OSPF Process 1 with Router ID 1.1.1.1
                      OSPF Protocol Information
 RouterID: 1.1.1.1         Border Router:
 Multi-VPN-Instance is not enabled
 Global DS-TE Mode: Non-Standard IETF Mode
 Graceful-restart capability: disabled
 Helper support capability  : not configured
 Applications Supported: MPLS Traffic-Engineering
```

```
Spf-schedule-interval: max 10000ms, start 500ms, hold 1000ms
Default ASE parameters: Metric: 1 Tag: 1 Type: 2
Route Preference: 10
ASE Route Preference: 150
SPF Computation Count: 13
RFC 1583 Compatible
Retransmission limitation is disabled
Area Count: 1   Nssa Area Count: 0
ExChange/Loading Neighbors: 0
Process total up interface count: 2
Process valid up interface count: 2


Area: 0.0.0.0          (MPLS TE not enabled)
Authtype: None   Area flag: Normal
SPF scheduled Count: 13
ExChange/Loading Neighbors: 0
Router ID conflict state: Normal
Area interface up count: 2


Interface: 192.168.1.254 (GigabitEthernet0/0/0)
Cost: 1       State: DR      Type: Broadcast    MTU: 1500
Priority: 1
Designated Router: 192.168.1.254
Backup Designated Router: 0.0.0.0
Timers: Hello 10 , Dead 40 , Poll 120 , Retransmit 5 , Transmit Delay 1

Interface: 192.168.2.1 (GigabitEthernet0/0/1)
Cost: 1       State: DR      Type: Broadcast    MTU: 1500
Priority: 1
Designated Router: 192.168.2.1
Backup Designated Router: 192.168.2.2
Timers: Hello 10 , Dead 40 , Poll 120 , Retransmit 5 , Transmit Delay 1
```

（7）查看路由器 OSPF 数据库信息，以 AR1 为例，结果如下：

```
[AR1]display ospf lsdb


 OSPF Process 1 with Router ID 1.1.1.1
        Link State Database
```

```
           Area: 0.0.0.0
 Type       LinkState ID    AdvRouter      Age    Len   Sequence     Metric
//路由器 LSA，只在本区域泛洪，不穿越 ABR。
 Router     2.2.2.2         2.2.2.2        635    48    80000008     1
 Router     1.1.1.1         1.1.1.1        753    48    80000007     1
 Router     5.5.5.5         5.5.5.5        460    48    80000005     1
 Router     3.3.3.3         3.3.3.3        499    48    80000008     1
//网络 LSA，由 DR 产生，只在选举 DR/BDR 的 Broadcast 和 NBMA 网络中才有，只在本区域泛洪，
//不穿越 ABR。
 Network    192.168.3.2     2.2.2.2        635    32    80000002     0
 Network    192.168.4.3     3.3.3.3        499    32    80000002     0
 Network    192.168.2.1     1.1.1.1        753    32    80000002     0
```

内容意义如下。

① LinkState ID：当 Type 为 Router 时为路由器 ID 号，代表路由器，而不是某一条链路。当 Type 为 Network 时为 DR 接口的 IP 地址。

② Adv Router：通告路由器的 ID 号。

③ Age：老化时间。

④ Len：LSA 的大小。

⑤ Sequence：LSA 序列号（来自 LSA 报头）。

⑥ Metric：度量值。

（8）查看接口 OSPF 信息，以 AR1 为例。

该命令主要用来查看所有接口的关于 OSPF 的信息，包括所在区域、OSPF 进程号、网络类型、代价、路由通告的统计信息、路由器 ID 号、DR 和 BDR 等。结果如下：

```
[AR1]display ospf interface

              OSPF Process 1 with Router ID 1.1.1.1
                         Interfaces

 Area: 0.0.0.0           (MPLS TE not enabled)
 IP Address      Type        State    Cost  Pri    DR              BDR
 192.168.1.254   Broadcast   DR       1     1      192.168.1.254   0.0.0.0
 192.168.2.1     Broadcast   DR       1     1      192.168.2.1     192.168.2.2
```

（9）查看 OSPF 邻居信息，以 AR1 为例，结果如下：

```
[AR1]display ospf peer brief

              OSPF Process 1 with Router ID 1.1.1.1
```

```
                  Peer Statistic Information
 ---------------------------------------------------------------
 Area Id          Interface                   Neighbor id      State
 0.0.0.0          GigabitEthernet0/0/1        2.2.2.2          Full
 ---------------------------------------------------------------
```

查看路由器的 OSPF 邻居信息是调试和排除 OSPF 故障的常用命令之一。

5. 多区域 OSPF 路由配置实验步骤

（1）布置拓扑。如图 4-19 所示，四台路由器连接了五个网段，其中，192.168.3.0/24 属于 area 0，192.168.2.0/24 和 192.168.1.0/24 属于 area 18，192.168.4.0/24 和 192.168.5.0/24 属于 area 36，其 IP 地址配置如表 4-10 所示。

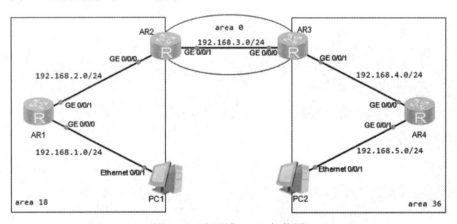

图 4-19　多区域 OSPF 拓扑图

表 4-10　各设备的 IP 地址配置（2）

设备名称	接口	IP 地址	默认网关	区域（area）
AR1	GE0/0/0	192.168.1.254/24		18
	GE0/0/1	192.168.2.1/24		18
AR2	GE0/0/0	192.168.2.2/24		18
	GE0/0/1	192.168.3.2/24		0
AR3	GE0/0/0	192.168.3.3/24		0
	GE0/0/1	192.168.4.3/24		36
AR4	GE0/0/0	192.168.4.4/24		36
	GE0/0/1	192.168.5.254/24		36
PC1	Ethernet0/0/1	192.168.1.1/24	192.168.1.254	
PC2	Ethernet0/0/1	192.168.5.1/24	192.168.5.254	

（2）配置路由器的 OSPF 路由，将全部路由器接口配置到 area 0。
AR1 中 OSPF 路由配置如下：

```
[AR1]ospf 1 router-id 1.1.1.1
```

```
[AR1-ospf-1]area 18
[AR1-ospf-1-area-0.0.0.18]network 192.168.1.0 0.0.0.255
[AR1-ospf-1-area-0.0.0.18]network 192.168.2.0 0.0.0.255
```

AR2 中 OSPF 路由配置如下：

```
[AR2]ospf 1 router-id 2.2.2.2
[AR2-ospf-1]area 18
[AR2-ospf-1-area-0.0.0.18]network 192.168.2.0 0.0.0.255
[AR2-ospf-1]area 0
[AR2-ospf-1-area-0.0.0.0]network 192.168.3.0 0.0.0.255
```

AR3 中 OSPF 路由配置如下：

```
[AR3]ospf 1 router-id 3.3.3.3
[AR3-ospf-1]area 0
[AR3-ospf-1-area-0.0.0.0]network 192.168.3.0 0.0.0.255
[AR3-ospf-1]area 36
[AR3-ospf-1-area-0.0.0.36]network 192.168.4.0 0.0.0.255
```

AR4 中 OSPF 路由配置如下：

```
[AR4]ospf 1 router-id 4.4.4.4
[AR4-ospf-1]area 36
[AR4-ospf-1-area-0.0.0.36]network 192.168.4.0 0.0.0.255
[AR4-ospf-1-area-0.0.0.36]network 192.168.5.0 0.0.0.255
```

（3）验证路由可达。由 PC1 ping PC2，如图 4-20 所示，结果是通的，说明 OSPF 路由配置正确。

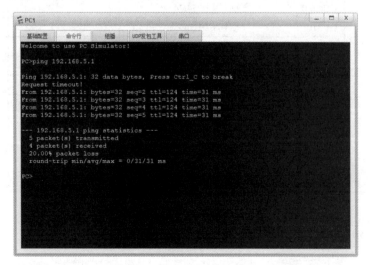

图 4-20　验证路由可达（2）

（4）查看路由器的路由表，以 AR1 为例，结果如下：

```
[AR1]display ip routing-table
Route Flags: R - relay, D - download to fib
------------------------------------------------------------
Routing Tables: Public
         Destinations : 13       Routes : 13
Destination/Mask      Proto   Pre  Cost  Flags  NextHop        Interface
        127.0.0.0/8   Direct  0    0     D      127.0.0.1      InLoopBack0
        127.0.0.1/32  Direct  0    0     D      127.0.0.1      InLoopBack0
127.255.255.255/32    Direct  0    0     D      127.0.0.1      InLoopBack0
      192.168.1.0/24  Direct  0    0     D      192.168.1.254  GigabitEthernet0/0/0
    192.168.1.254/32  Direct  0    0     D      127.0.0.1      GigabitEthernet0/0/0
    192.168.1.255/32  Direct  0    0     D      127.0.0.1      GigabitEthernet0/0/0
      192.168.2.0/24  Direct  0    0     D      192.168.2.1    GigabitEthernet0/0/1
      192.168.2.1/32  Direct  0    0     D      127.0.0.1      GigabitEthernet0/0/1
    192.168.2.255/32  Direct  0    0     D      127.0.0.1      GigabitEthernet0/0/1
      192.168.3.0/24  OSPF    10   2     D      192.168.2.2    GigabitEthernet0/0/1
      192.168.4.0/24  OSPF    10   3     D      192.168.2.2    GigabitEthernet0/0/1
      192.168.5.0/24  OSPF    10   4     D      192.168.2.2    GigabitEthernet0/0/1
  255.255.255.255/32  Direct  0    0     D      127.0.0.1      InLoopBack0
```

通过查看路由表可知，AR1 中有五条路由，对应拓扑中的五个网段，其中 192.168.1.0/24 和 192.168.2.0/24 为路由器的直连路由，192.168.3.0/24、192.168.4.0/24、192.168.5.0/24 为 OSPF 路由。

其他路由器的路由表项，请读者自行分析。

（5）查看 OSPF 中各区域邻居的信息，以 AR1 为例，结果如下：

```
[AR1]display ospf peer
            OSPF Process 1 with Router ID 1.1.1.1
                    Neighbors
Area 0.0.0.18 interface 192.168.2.1(GigabitEthernet0/0/1)'s neighbors
Router ID: 2.2.2.2       Address: 192.168.2.2
  State: Full Mode:Nbr is Master Priority: 1
  DR: 192.168.2.1 BDR: 192.168.2.2 MTU: 0
  Dead timer due in 33 sec
  Retrans timer interval: 5
  Neighbor is up for 00:16:54
Authentication Sequence: [ 0 ]
```

（6）查看 OSPF 进程及区域细节的数据，以 AR1 为例，结果如下：

```
[AR1]display ospf brief
              OSPF Process 1 with Router ID 1.1.1.1
                    OSPF Protocol Information

RouterID: 1.1.1.1         Border Router:
Multi-VPN-Instance is not enabled
Global DS-TE Mode: Non-Standard IETF Mode
Graceful-restart capability: disabled
Helper support capability : not configured
Applications Supported: MPLS Traffic-Engineering
Spf-schedule-interval: max 10000ms, start 500ms, hold 1000ms
Default ASE parameters: Metric: 1 Tag: 1 Type: 2
Route Preference: 10
ASE Route Preference: 150
SPF Computation Count: 7
RFC 1583 Compatible
Retransmission limitation is disabled
Area Count: 1  Nssa Area Count: 0
ExChange/Loading Neighbors: 0
Process total up interface count: 2
Process valid up interface count: 2

Area: 0.0.0.18          (MPLS TE not enabled)
Authtype: None   Area flag: Normal
SPF scheduled Count: 7
ExChange/Loading Neighbors: 0
Router ID conflict state: Normal
Area interface up count: 2

Interface: 192.168.1.254 (GigabitEthernet0/0/0)
Cost: 1      State: DR      Type: Broadcast    MTU: 1500
Priority: 1
Designated Router: 192.168.1.254
Backup Designated Router: 0.0.0.0
Timers: Hello 10 , Dead 40 , Poll 120 , Retransmit 5 , Transmit Delay 1

Interface: 192.168.2.1 (GigabitEthernet0/0/1)
Cost: 1      State: DR      Type: Broadcast    MTU: 1500
```

```
   Priority: 1
   Designated Router: 192.168.2.1
   Backup Designated Router: 192.168.2.2
   Timers: Hello 10 , Dead 40 , Poll 120 , Retransmit 5 , Transmit Delay 1
```

(7) 查看路由器 OSPF 数据库信息，以 AR1 为例，结果如下：

```
[AR1]display ospf lsdb
            OSPF Process 1 with Router ID 1.1.1.1
                    Link State Database
        Area: 0.0.0.18
 Type       LinkState ID      AdvRouter         Age   Len   Sequence    Metric
 Router     2.2.2.2           2.2.2.2           1150  36    80000004    1
 Router     1.1.1.1           1.1.1.1           1147  48    80000007    1
 Network    192.168.2.1       1.1.1.1           1147  32    80000002    0
//由其他区域泛洪过来而得到的汇总链路（三类LSA）。
 Sum-Net    192.168.5.0       2.2.2.2           884   28    80000001    3
 Sum-Net    192.168.4.0       2.2.2.2           1007  28    80000001    2
 Sum-Net    192.168.3.0       2.2.2.2           1143  28    80000001    1
```

内容意义如下。

① LinkState ID：当 Type 为 Sum-Net 时，代表汇总过来的网络。

② AdvRouter：通告路由器的 ID 号。

③ Age：老化时间。

④ Len：LSA 的大小。

⑤ Sequence：LSA 序列号（来自 LSA 报头）。

⑥ Metric：度量值。

(8) 查看接口 OSPF 信息，以 AR1 为例。

该命令主要用来查看所有接口的关于 OSPF 的信息，包括所在区域、OSPF 进程号、网络类型、代价、路由通告的统计信息、路由器 ID 号、DR 和 BDR 等。结果如下：

```
[AR1]display ospf interface
            OSPF Process 1 with Router ID 1.1.1.1
                    Interfaces
 Area: 0.0.0.18          (MPLS TE not enabled)
 IP Address       Type       State  Cost  Pri    DR              BDR
 192.168.1.254    Broadcast  DR     1     1      192.168.1.254   0.0.0.0
 192.168.2.1      Broadcast  DR     1     1      192.168.2.1     192.168.2.2
```

(9) 查看 OSPF 邻居信息，以 AR1 为例，结果如下：

```
[AR1]display ospf peer brief

            OSPF Process 1 with Router ID 1.1.1.1
                    Peer Statistic Information
 -----------------------------------------------------------------
 Area Id           Interface                 Neighbor id     State
 0.0.0.18          GigabitEthernet0/0/1      2.2.2.2         Full
 -----------------------------------------------------------------
```

查看路由器的 OSPF 邻居信息是调试和排除 OSPF 故障的常用命令之一。

（10）调试 OSPF 事件，主要包括显示发送/接收 hello 包、邻居改变事件、DR 选取、如何建立邻接关系等。结果如下：

```
<AR1>terminal monitor
Info: Current terminal monitor is on.
<AR1>terminal debugging
Info: Current terminal debugging is on.
<AR1>debugging ospf packet hello interface GigabitEthernet 0/0/1
Oct 21 2020 18:09:58.58.1-08:00 AR1 RM/6/RMDEBUG:
 FileID: 0xd0178025 Line: 559 Level: 0x20
 OSPF 1: SEND Packet. Interface: GigabitEthernet0/0/1
Source Address: 192.168.2.1
Destination Address: 224.0.0.5
Ver# 2, Type: 1 (Hello)
Length: 48, Router: 1.1.1.1
Area: 0.0.0.18, Chksum: 712eAuType: 00
Key(ascii): * * * * * * * *
Net Mask: 255.255.255.0
Hello Int: 10, Option: _E_
//以太网或者点对点网络默认发送 hello 包的时间间隔是 10s，即每隔 10s 发送 hello 包。不同的
//网络类型发送 hello 包的频率不一样，如果是非广播多路访问网络（NBMA 网络），那么发送 hello
//包的时间间隔是 30s。当然，这个时间都可以使用命令修改。
Rtr Priority: 1, Dead Int: 40
DR: 192.168.2.1
BDR: 192.168.2.2
# Attached Neighbors: 1
Neighbor: 2.2.2.2
//这时，在 AR2 上 shutdown 与 AR1 的接口（GE0/0/0），当达到死亡时间后，AR1 认为邻接关系
//断掉，由于是广播型网络，因此拓扑改变后会重新选取 DR 和 BDR，但此时该区域中实际上已经没
```

```
//有网络了，结果如下：
<AR1>
Oct  21  2020  18:13:31-08:00  AR1  %%01IFPDT/4/IF_STATE(l)[0]:Interface
GigabitEthernet0/0/1 has turned into DOWN state.
<AR1>
Oct 21 2020 18:13:31-08:00 AR1 %%01IFNET/4/LINK_STATE(l)[1]:The line protocol
IP on the interface GigabitEthernet0/0/1 has entered the DOWN state.
<AR1>
Oct  21  2020  18:13:31-08:00  AR1  %%01OSPF/3/NBR_CHG_DOWN(l)[2]:Neighbor
event:neighbor state changed to Down. (ProcessId=256,NeighborAddress=2.2.2.2,
NeighborEvent=KillNbr,NeighborPreviousState=Full,NeighborCurrentState=Down)
<AR1>
Oct  21  2020  18:13:31-08:00  AR1  %%01OSPF/3/NBR_DOWN_REASON(l)[3]:Neighbor
state leaves full or changed to Down. (ProcessId=256, NeighborRouterId=2.2.2.2,
NeighborAreaId=301989888,
NeighborInterface=GigabitEthernet0/0/1,NeighborDownImmediate
reason=Neighbor Down Due to Kill Neighbor, NeighborDownPrimeReason=Physical
Interface State Change, NeighborChangeTime=2020-10-21 18:13:31-08:00)
```

实验六：外部网关协议 BGP 实验

1. 实验目的

（1）理解 BGP 的含义。
（2）掌握 BGP 的配置方法。

2. 基础知识

边界网关协议（Border Gateway Protocol，BGP）是一种路径向量协议，属于外部网关协议。BGP 用于在不同自治系统（AS）之间选择路由，力求寻找一条能够到达目的网络且比较好的路由，而不是寻找一条最佳路由。BGP 有以下特点。

（1）BGP 分为 eBGP（外部 BGP）和 iBGP（内部 BGP），两个 AS 之间的发言人运行 eBGP，而在一个 AS 内部的网络则运行 iBGP，它们使用同样的报文格式和属性类型。eBGP 发言人将获得的 eBGP 路由通过 iBGP 传递给 AS 内部的 iBGP 邻居，再由其进一步转化为自己的路由表项目。

（2）BGP 邻居间进行通信是建立在 TCP 连接之上的。TCP 连接是建立在 IP 协议之上的一种一对一的通信，这意味着 BGP 的邻居并非物理连接上的邻居，而是可以经过 IP 路由后到达的邻居。

（3）一系列的 AS 路径和经过这些 AS 所能到达的目的地组成了 BGP 的世界，其通过路径向量路由选择协议选出最合适的路由。

详细解释请参考教材《计算机网络（第 8 版）》4.6.4 节。

本实验配置命令如表 4-11 所示。

表 4-11 本实验配置命令

命令格式	命令说明
bgp as-number	启动 BGP，进入 BGP 视图
router-id X.X.X.X	配置 Router ID
ipv4-family unicast	进入单播地址族视图
import-route ospf 1	将 OSPF 路由引入 BGP
peer X.X.X.X as-number XXX	创建对等体
peer X.X.X.X connect-interface 接口	指定发送 BGP 报文的源接口，指定发起连接时使用的源地址
display bgp routing-table	查看 BGP 路由表
display bgp peer	查看 BGP 邻居的摘要信息
display bgp peer verbose	查看 BGP 邻居的详细信息

3. 实验流程

本实验配置 BGP 路由，要求各 IP 地址全部可达。实验流程如图 4-21 所示。

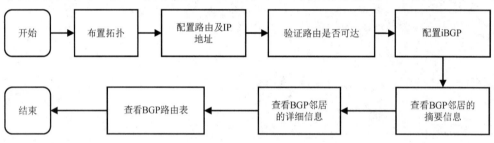

图 4-21 实验流程图

4. iBGP 配置实验步骤

（1）按图 4-22 所示布置拓扑，并按表 4-12 配置各设备的 IP 地址。

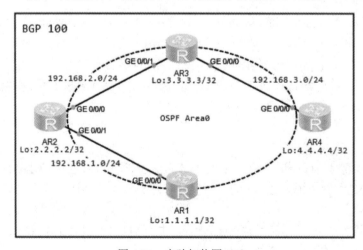

图 4-22 实验拓扑图（1）

表 4-12 各设备的 IP 地址配置（1）

设备名称	接口	IP 地址	LoopBack0接口IP地址	区域（area）	BGP
AR1	GE0/0/0	192.168.1.1/24	1.1.1.1/32	0	100
AR2	GE0/0/0	192.168.2.2/24	2.2.2.2/32	0	100
	GE0/0/1	192.168.1.2/24	2.2.2.2/32	0	100
AR3	GE0/0/0	192.168.3.3/24	3.3.3.3/32	0	100
	GE0/0/1	192.168.2.3/24	3.3.3.3/32	0	100
AR4	GE0/0/0	192.168.3.4/24	4.4.4.4/32	0	100

（2）配置路由。

配置 AR1 的路由。配置如下：

```
[Huawei]sy AR1
[AR1]int LoopBack 0
[AR1-LoopBack0]ip address 1.1.1.1 255.255.255.255
[AR1-LoopBack0]int g0/0/0
[AR1-GigabitEthernet0/0/0]ip address 192.168.1.1 255.255.255.0
[AR1-GigabitEthernet0/0/0]q
[AR1]ospf 1 router-id 1.1.1.1
[AR1-ospf-1]area 0.0.0.0
[AR1-ospf-1-area-0.0.0.0]network 1.1.1.1 0.0.0.0
[AR1-ospf-1-area-0.0.0.0]network 192.168.1.1 0.0.0.0
[AR1-ospf-1-area-0.0.0.0]q
[AR1-ospf-1]q
```

配置 AR2 的路由。配置如下：

```
[Huawei]sy AR2
[AR2]int LoopBack 0
[AR2-LoopBack0]ip address 2.2.2.2 255.255.255.255
[AR2-LoopBack0]int g0/0/0
[AR2-GigabitEthernet0/0/0]ip address 192.168.2.2 255.255.255.0
[AR2-GigabitEthernet0/0/0]int g0/0/1
[AR2-GigabitEthernet0/0/1]ip address 192.168.1.2 255.255.255.0
[AR2-GigabitEthernet0/0/1]q
[AR2]ospf 1 router-id 2.2.2.2
[AR2-ospf-1]area 0.0.0.0
[AR2-ospf-1-area-0.0.0.0]network 2.2.2.2 0.0.0.0
[AR2-ospf-1-area-0.0.0.0]network 192.168.1.2 0.0.0.0
[AR2-ospf-1-area-0.0.0.0]network 192.168.2.2 0.0.0.0
```

```
[AR2-ospf-1-area-0.0.0.0]q
[AR2-ospf-1]q
```

配置 AR3 的路由。配置如下：

```
[Huawei]sy AR3
[AR3]int LoopBack 0
[AR3-LoopBack0]ip address 3.3.3.3 255.255.255.255
[AR3-LoopBack0]int g0/0/0
[AR3-GigabitEthernet0/0/0]ip address 192.168.3.3 255.255.255.0
[AR3-GigabitEthernet0/0/0]int g0/0/1
[AR3-GigabitEthernet0/0/1]ip address 192.168.2.3 255.255.255.0
[AR3-GigabitEthernet0/0/1]q
[AR3]ospf 1 router-id 3.3.3.3
[AR3-ospf-1]area 0.0.0.0
[AR3-ospf-1-area-0.0.0.0]network 3.3.3.3 0.0.0.0
[AR3-ospf-1-area-0.0.0.0]network 192.168.2.3 0.0.0.0
[AR3-ospf-1-area-0.0.0.0]network 192.168.3.3 0.0.0.0
[AR3-ospf-1-area-0.0.0.0]q
[AR3-ospf-1]q
```

配置 AR4 的路由。配置如下：

```
[Huawei]sy AR4
[AR4]int LoopBack 0
[AR4-LoopBack0]ip address 4.4.4.4 255.255.255.255
[AR4-LoopBack0]int g0/0/0
[AR4-GigabitEthernet0/0/0]ip address 192.168.3.4 255.255.255.0
[AR4-GigabitEthernet0/0/0]q
[AR4]ospf 1 router-id 4.4.4.4
[AR4-ospf-1]area 0.0.0.0
[AR4-ospf-1-area-0.0.0.0]network 4.4.4.4 0.0.0.0
[AR4-ospf-1-area-0.0.0.0]network 192.168.3.4 0.0.0.0
[AR4-ospf-1-area-0.0.0.0]q
[AR4-ospf-1]q
```

（3）配置 iBGP 连接。

AR1 的配置如下：

```
[AR1]bgp 100
[AR1-bgp]router-id 1.1.1.1
```

```
[AR1-bgp]peer 2.2.2.2 as-number 100
[AR1-bgp]peer 2.2.2.2 connect-interface LoopBack0
[AR1-bgp]ipv4-family unicast
[AR1-bgp-af-ipv4]import-route ospf 1
```

AR2 的配置如下：

```
[AR2]bgp 100
[AR2-bgp]router-id 2.2.2.2
[AR2-bgp]peer 1.1.1.1 as-number 100
[AR2-bgp]peer 1.1.1.1 connect-interface LoopBack0
[AR2-bgp]peer 3.3.3.3 as-number 100
[AR2-bgp]peer 3.3.3.3 connect-interface LoopBack0
[AR2-bgp]ipv4-family unicast
[AR2-bgp-af-ipv4]import-route ospf 1
```

AR3 的配置如下：

```
[AR3]bgp 100
[AR3-bgp]router-id 3.3.3.3
[AR3-bgp]peer 2.2.2.2 as-number 100
[AR3-bgp]peer 2.2.2.2 connect-interface LoopBack0
[AR3-bgp]peer 4.4.4.4 as-number 100
[AR3-bgp]peer 4.4.4.4 connect-interface LoopBack0
[AR3-bgp]ipv4-family unicast
[AR3-bgp-af-ipv4]import-route ospf 1
```

AR4 的配置如下：

```
[AR4]bgp 100
[AR4-bgp]router-id 4.4.4.4
[AR4-bgp]peer 3.3.3.3 as-number 100
[AR4-bgp]peer 3.3.3.3 connect-interface LoopBack0
[AR4-bgp]ipv4-family unicast
[AR4-bgp-af-ipv4]import-route ospf 1
```

（4）显示 BGP 邻居的摘要信息，以 AR2 为例，结果如下：

```
[AR2]display bgp peer
 BGP local router ID : 2.2.2.2
 Local AS number : 100
 Total number of peers : 2         Peers in established state : 2
```

```
Peer        V    AS    MsgRcvd  MsgSent   OutQ  Up/Down    State       Pre  fRcv
1.1.1.1     4    100     57        57      0   00:48:38   Established       7
3.3.3.3     4    100     56        57      0   00:47:46   Established       7
```

（5）显示 BGP 邻居的详细信息，以 AR2 为例，结果如下：

```
[AR2]display bgp peer verbose

    BGP Peer is 1.1.1.1,  remote AS 100
    Type: IBGP link
    BGP version 4, Remote router ID 1.1.1.1
    Update-group ID: 1
    BGP current state: Established, Up for 00h50m07s
    BGP current event: KATimerExpired
    BGP last state: OpenConfirm
    BGP Peer Up count: 1
    Received total routes: 7
    Received active routes total: 0
    Advertised total routes: 7
    Port:  Local - 49276 Remote - 179
    Configured: Connect-retry Time: 32 sec
    Configured: Active Hold Time: 180 sec   Keepalive Time:60 sec
    Received  : Active Hold Time: 180 sec
    Negotiated: Active Hold Time: 180 sec   Keepalive Time:60 sec
    Peer optional capabilities:
    Peer supports bgp multi-protocol extension
    Peer supports bgp route refresh capability
    Peer supports bgp 4-byte-as capability
    Address family IPv4 Unicast: advertised and received
 Received: Total 59 messages
        Update messages      7
        Open messages        1
        KeepAlive messages      51
        Notification messages   0
        Refresh messages        0
 Sent: Total 59 messages
        Update messages      7
        Open messages        1
        KeepAlive messages      51
```

```
        Notification messages    0
        Refresh messages         0
Authentication type configured: None
Last keepalive received: 2021/03/12 11:15:30 UTC-08:00
Last keepalive sent    : 2021/03/12 11:15:31 UTC-08:00
Last update     received: 2021/03/12 10:27:33 UTC-08:00
Last update     sent   : 2021/03/12 10:27:47 UTC-08:00
Minimum route advertisement interval is 15 seconds
Optional capabilities:
Route refresh capability has been enabled
4-byte-as capability has been enabled
Connect-interface has been configured
Peer Preferred Value: 0
Routing policy configured:
No routing policy is configured


   BGP Peer is 3.3.3.3,  remote AS 100
   Type: IBGP link
   BGP version 4, Remote router ID 3.3.3.3
   Update-group ID: 1
   BGP current state: Established, Up for 00h49m15s
   BGP current event: RecvKeepalive
   BGP last state: OpenConfirm
   BGP Peer Up count: 1
   Received total routes: 7
   Received active routes total: 0
   Advertised total routes: 7
   Port: Local - 179   Remote - 50930
   Configured: Connect-retry Time: 32 sec
   Configured: Active Hold Time: 180 sec   Keepalive Time:60 sec
   Received : Active Hold Time: 180 sec
   Negotiated: Active Hold Time: 180 sec   Keepalive Time:60 sec
   Peer optional capabilities:
   Peer supports bgp multi-protocol extension
   Peer supports bgp route refresh capability
   Peer supports bgp 4-byte-as capability
   Address family IPv4 Unicast: advertised and received
Received: Total 58 messages
        Update messages         7
```

```
              Open messages                1
              KeepAlive messages          50
              Notification messages        0
              Refresh messages             0
Sent: Total 59 messages
              Update messages              7
              Open messages                2
              KeepAlive messages          50
              Notification messages        0
              Refresh messages             0
Authentication type configured: None
Last keepalive received: 2021/03/12 11:15:23 UTC-08:00
Last keepalive sent    : 2021/03/12 11:15:22 UTC-08:00
Last update   received: 2021/03/12 10:28:03 UTC-08:00
Last update     sent   : 2021/03/12 10:27:47 UTC-08:00
Minimum route advertisement interval is 15 seconds
Optional capabilities:
Route refresh capability has been enabled
4-byte-as capability has been enabled
Connect-interface has been configured
Peer Preferred Value: 0
Routing policy configured:
No routing policy is configured
```

（6）显示 BGP 路由表，以 AR2 为例，结果如下：

```
[AR2]display bgp routing-table
BGP Local router ID is 2.2.2.2
Status codes: * - valid, > - best, d - damped,
          h - history, i - internal, s - suppressed, S - Stale
          Origin : i - IGP, e - EGP, ? - incomplete
Total Number of Routes: 21
   Network            NextHop          MED      LocPrf    PrefVal  Path/Ogn
 *>  1.1.1.1/32       0.0.0.0          1                  0        ?
 *i                   3.3.3.3          2        100       0        ?
  i                   1.1.1.1          0        100       0        ?
 *>  2.2.2.2/32       0.0.0.0          0                  0        ?
 *i                   1.1.1.1          1        100       0        ?
 *i                   3.3.3.3          1        100       0        ?
```

*> 3.3.3.3/32	0.0.0.0	1		0	?
* i	1.1.1.1	2	100	0	?
i	3.3.3.3	0	100	0	?
*> 4.4.4.4/32	0.0.0.0	2		0	?
* i	3.3.3.3	1	100	0	?
* i	1.1.1.1	3	100	0	?
*> 192.168.1.0	0.0.0.0	0		0	?
* i	1.1.1.1	0	100	0	?
* i	3.3.3.3	2	100	0	?
*> 192.168.2.0	0.0.0.0	0		0	?
* i	3.3.3.3	0	100	0	?
* i	1.1.1.1	2	100	0	?
*> 192.168.3.0	0.0.0.0	2		0	?
* i	3.3.3.3	0	100	0	?
* i	1.1.1.1	3	100	0	?

（7）验证连通性即测试路由可达，由 AR1 ping AR4，结果如图 4-23 所示。

```
<AR1>ping 192.168.3.4
  PING 192.168.3.4: 56  data bytes, press CTRL_C to break
    Reply from 192.168.3.4: bytes=56 Sequence=1 ttl=253 time=30 ms
    Reply from 192.168.3.4: bytes=56 Sequence=2 ttl=253 time=30 ms
    Reply from 192.168.3.4: bytes=56 Sequence=3 ttl=253 time=30 ms
    Reply from 192.168.3.4: bytes=56 Sequence=4 ttl=253 time=20 ms
    Reply from 192.168.3.4: bytes=56 Sequence=5 ttl=253 time=20 ms

  --- 192.168.3.4 ping statistics ---
    5 packet(s) transmitted
    5 packet(s) received
    0.00% packet loss
    round-trip min/avg/max = 20/26/30 ms
```

图 4-23 测试路由可达（1）

5. eBGP 配置实验步骤

（1）按图 4-24 所示布置拓扑，并按表 4-13 配置各设备的 IP 地址。

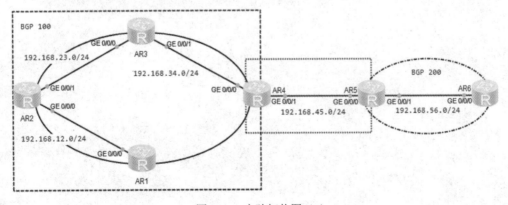

图 4-24 实验拓扑图（2）

表 4-13　各设备的 IP 地址配置（2）

设备名称	接口	IP 地址	LoopBack0 接口 IP 地址	OSPF 进程	区域（area）	BGP
AR1	GE0/0/0	192.168.12.1/24	1.1.1.1/32	1	0	100
AR2	GE0/0/0	192.168.12.2/24	2.2.2.2/32	1	0	100
	GE0/0/1	192.168.23.2/24		1	0	100
AR3	GE0/0/0	192.168.23.3/24	3.3.3.3/32	1	0	100
	GE0/0/1	192.168.34.3/24		1	0	100
AR4	GE0/0/0	192.168.34.4/24	4.4.4.4/32	1	0	100
	GE0/0/1	192.168.45.4/24		2	0	100
AR5	GE0/0/0	192.168.45.5/24	5.5.5.5/32	2	0	200
	GE0/0/1	192.168.56.5/24		3	0	200
AR6	GE0/0/0	192.168.56.6/24	6.6.6.6/32	3	0	200

（2）配置路由。

配置 AR1 的路由。配置如下：

```
[Huawei]sy AR1
[AR1]int LoopBack 0
[AR1-LoopBack0]ip address 1.1.1.1 32
[AR1-LoopBack0]q
[AR1]int g0/0/0
[AR1-GigabitEthernet0/0/0]ip add 192.168.12.1 24
[AR1-GigabitEthernet0/0/0]q
[AR1]ospf 1 router-id 1.1.1.1
[AR1-ospf-1]area 0
[AR1-ospf-1-area-0.0.0.0]network 192.168.12.1 0.0.0.0
[AR1-ospf-1-area-0.0.0.0]network 1.1.1.1 0.0.0.0
```

配置 AR2 的路由。配置如下：

```
[Huawei]sy AR2
[AR2]int LoopBack 0
[AR2-LoopBack0]ip address 2.2.2.2 32
[AR2-LoopBack0]q
[AR2]int g0/0/0
[AR2-GigabitEthernet0/0/0]ip add 192.168.12.2 24
[AR2-GigabitEthernet0/0/0]int g0/0/1
[AR2-GigabitEthernet0/0/1]ip add 192.168.23.2 24
[AR2-GigabitEthernet0/0/1]q
[AR2]ospf 1 router-id 2.2.2.2
[AR2-ospf-1]area 0
```

```
[AR2-ospf-1-area-0.0.0.0]network 192.168.12.2 0.0.0.0
[AR2-ospf-1-area-0.0.0.0]network 192.168.23.2 0.0.0.0
[AR2-ospf-1-area-0.0.0.0]network 2.2.2.2 0.0.0.0
```

配置 AR3 的路由。配置如下：

```
[Huawei]sy AR3
[AR3]int LoopBack 0
[AR3-LoopBack0]ip address 3.3.3.3 32
[AR3-LoopBack0]q
[AR3]int g0/0/0
[AR3-GigabitEthernet0/0/0]ip add 192.168.23.3 24
[AR3-GigabitEthernet0/0/0]int g0/0/1
[AR3-GigabitEthernet0/0/1]ip add 192.168.34.3 24
[AR3-GigabitEthernet0/0/1]q
[AR3]ospf 1 router-id 3.3.3.3
[AR3-ospf-1]area 0
[AR3-ospf-1-area-0.0.0.0]network 192.168.23.3 0.0.0.0
[AR3-ospf-1-area-0.0.0.0]network 192.168.34.3 0.0.0.0
[AR3-ospf-1-area-0.0.0.0]network 3.3.3.3 0.0.0.0
```

配置 AR4 的路由。配置如下：

```
[Huawei]sy AR4
[AR4]int LoopBack 0
[AR4-LoopBack0]ip address 4.4.4.4 32
[AR4-LoopBack0]q
[AR4]int g0/0/0
[AR4-GigabitEthernet0/0/0]ip add 192.168.34.4 24
[AR4-GigabitEthernet0/0/0]int g0/0/1
[AR4-GigabitEthernet0/0/1]ip add 192.168.45.4 24
[AR4-GigabitEthernet0/0/1]q
[AR4]ospf 1 router-id 4.4.4.4
[AR4-ospf-1]import-route ospf 2
[AR4-ospf-1]area 0
[AR4-ospf-1-area-0.0.0.0]network 192.168.34.4 0.0.0.0
[AR4-ospf-1-area-0.0.0.0]network 4.4.4.4 0.0.0.0
[AR4-ospf-1-area-0.0.0.0]q
[AR4-ospf-1]q
[AR4]ospf 2 router-id 4.4.4.4
```

```
[AR4-ospf-2]area0
[AR4-ospf-2]importospf 1
[AR4-ospf-2-area-0.0.0.0]network 192.168.45.4 0.0.0.0
```

配置 AR5 的路由。配置如下：

```
[Huawei]sy AR5
[AR5]int LoopBack 0
[AR5-LoopBack0]ip address 5.5.5.5 32
[AR5-LoopBack0]q
[AR5]int g0/0/0
[AR5-GigabitEthernet0/0/0]ip add 192.168.45.5 24
[AR5-GigabitEthernet0/0/0]int g0/0/1
[AR5-GigabitEthernet0/0/1]ip add 192.168.56.5 24
[AR5-GigabitEthernet0/0/1]q
[AR5]ospf 2 router-id 5.5.5.5
[AR5-ospf-2]importospf 3
[AR5-ospf-2]area 0
[AR5-ospf-2-area-0.0.0.0]network 192.168.45.5 0.0.0.0
[AR5-ospf-2-area-0.0.0.0]network 5.5.5.5 0.0.0.0
[AR5-ospf-2-area-0.0.0.0]q
[AR5-ospf-2]q
[AR5]ospf 3 router-id 5.5.5.5
[AR5-ospf-3]importospf 2
[AR5-ospf-3]area 0
[AR5-ospf-3-area-0.0.0.0]network 192.168.56.5 0.0.0.0
```

配置 AR6 的路由。配置如下：

```
[Huawei]sy AR6
[AR6]int LoopBack 0
[AR6-LoopBack0]ip address 6.6.6.6 32
[AR6-LoopBack0]q
[AR6]int g0/0/0
[AR6-GigabitEthernet0/0/0]ip add 192.168.56.6 24
[AR6-GigabitEthernet0/0/0]q
[AR6]ospf 3 router-id 6.6.6.6
[AR6-ospf-3]area 0
[AR6-ospf-3-area-0.0.0.0]network 192.168.56.6 0.0.0.0
[AR6-ospf-3-area-0.0.0.0]network 6.6.6.6 0.0.0.0
```

(3) 配置 eBGP 连接。
AR1 的配置如下:

```
[AR1]bgp 100
[AR1-bgp]router-id 1.1.1.1
[AR1-bgp]peer 2.2.2.2 as-number 100
[AR1-bgp]peer 2.2.2.2 connect-interface LoopBack 0
```

AR2 的配置如下:

```
[AR2]bgp 100
[AR2-bgp]router-id 2.2.2.2
[AR2-bgp]peer 1.1.1.1 as-number 100
[AR2-bgp]peer 1.1.1.1 connect-interface LoopBack 0
[AR2-bgp]peer 3.3.3.3 as-number 100
[AR2-bgp]peer 3.3.3.3 connect-interface LoopBack 0
```

AR3 的配置如下:

```
[AR3]bgp 100
[AR3-bgp]router-id 3.3.3.3
[AR3-bgp]peer 2.2.2.2 as-number 100
[AR3-bgp]peer 2.2.2.2 connect-interface LoopBack 0
[AR3-bgp]peer 4.4.4.4 as-number 100
[AR3-bgp]peer 4.4.4.4 connect-interface LoopBack 0
```

AR4 的配置如下:

```
[AR4]bgp 100
[AR4-bgp]router-id 4.4.4.4
[AR4-bgp]peer 3.3.3.3 as-number 100
[AR4-bgp]peer 3.3.3.3 connect-interface LoopBack 0
[AR4-bgp]peer 5.5.5.5 as-number 200
[AR4-bgp]peer 5.5.5.5 connect-interface LoopBack 0
[AR4-bgp]peer 5.5.5.5 ebgp-max-hop 2
[AR4-bgp]ipv4-family unicast
[AR4-bgp-af-ipv4]import-route ospf 1
[AR4-bgp-af-ipv4]import-route ospf 2
```

AR5 的配置如下:

```
[AR5]bgp 200
```

```
[AR5-bgp]router-id 5.5.5.5
[AR5-bgp]peer 4.4.4.4 as-number 100
[AR5-bgp]peer 4.4.4.4 connect-interface LoopBack 0
[AR5-bgp]peer 4.4.4.4 ebgp-max-hop 2
[AR5-bgp]peer 6.6.6.6 as-number 200
[AR5-bgp]peer 6.6.6.6 connect-interface LoopBack 0
[AR5-bgp]ipv4-family unicast
[AR5-bgp-af-ipv4]import-route ospf 2
[AR5-bgp-af-ipv4]import-route ospf 3
```

AR6 的配置如下：

```
[AR6]bgp 200
[AR6-bgp]router-id 6.6.6.6
[AR6-bgp]peer 5.5.5.5 as-number 100
[AR6-bgp]peer 5.5.5.5 connect-interface LoopBack 0
```

（4）显示 BGP 邻居的摘要信息，以 AR4 为例，结果如下：

```
[AR4]display bgp peer
 BGP local router ID : 4.4.4.4
 Local AS number : 100
 Total number of peers : 2       Peers in established state : 2
  Peer            V     AS   MsgRcvd  MsgSent  OutQ  Up/Down   State  PrefRcv
  3.3.3.3         4    100      9       19      0   00:07:02 Established  0
  5.5.5.5         4    200     13       15      0   00:06:36 Established  11
```

（5）显示 BGP 邻居的详细信息，以 AR4 为例，结果如下：

```
[AR4]display bgp peer verbose

    BGP Peer is 3.3.3.3,  remote AS 100
    Type: IBGP link
    BGP version 4, Remote router ID 3.3.3.3
    Update-group ID: 2
    BGP current state: Established, Up for 00h08m50s
    BGP current event: RecvKeepalive
    BGP last state: OpenConfirm
    BGP Peer Up count: 1
    Received total routes: 0
    Received active routes total: 0
```

```
    Advertised total routes: 11
    Port:  Local - 179   Remote - 50482
    Configured: Connect-retry Time: 32 sec
    Configured: Active Hold Time: 180 sec    Keepalive Time:60 sec
    Received  : Active Hold Time: 180 sec
    Negotiated: Active Hold Time: 180 sec    Keepalive Time:60 sec
    Peer optional capabilities:
    Peer supports bgp multi-protocol extension
    Peer supports bgp route refresh capability
    Peer supports bgp 4-byte-as capability
    Address family IPv4 Unicast: advertised and received
Received: Total 10 messages
        Update messages         0
        Open messages           1
        KeepAlive messages      9
        Notification messages   0
        Refresh messages        0
Sent: Total 20 messages
        Update messages         9
        Open messages           2
        KeepAlive messages      9
        Notification messages   0
        Refresh messages        0
Authentication type configured: None
Last keepalive received: 2021/03/15 19:22:05 UTC-08:00
Last keepalive sent    : 2021/03/15 19:22:04 UTC-08:00
Last update    sent    : 2021/03/15 19:14:29 UTC-08:00
Minimum route advertisement interval is 15 seconds
Optional capabilities:
Route refresh capability has been enabled
4-byte-as capability has been enabled
Connect-interface has been configured
Peer Preferred Value: 0
Routing policy configured:
No routing policy is configured

    BGP Peer is 5.5.5.5,  remote AS 200
    Type: EBGP link
    BGP version 4, Remote router ID 5.5.5.5
```

```
    Update-group ID: 0
    BGP current state: Established, Up for 00h08m25s
    BGP current event: RecvKeepalive
    BGP last state: OpenConfirm
    BGP Peer Up count: 1
    Received total routes: 11
    Received active routes total: 0
    Advertised total routes: 11
    Port:  Local - 50718 Remote - 179
    Configured: Connect-retry Time: 32 sec
    Configured: Active Hold Time: 180 sec   Keepalive Time:60 sec
    Received  : Active Hold Time: 180 sec
    Negotiated: Active Hold Time: 180 sec   Keepalive Time:60 sec
    Peer optional capabilities:
    Peer supports bgp multi-protocol extension
    Peer supports bgp route refresh capability
    Peer supports bgp 4-byte-as capability
    Address family IPv4 Unicast: advertised and received
Received: Total 15 messages
        Update messages          5
        Open messages            1
        KeepAlive messages       9
        Notification messages    0
        Refresh messages         0
Sent: Total 17 messages
        Update messages          7
        Open messages            1
        KeepAlive messages       9
        Notification messages    0
        Refresh messages         0
Authentication type configured: None
Last keepalive received: 2021/03/15 19:22:31 UTC-08:00
Last keepalive sent    : 2021/03/15 19:22:30 UTC-08:00
Last update    received: 2021/03/15 19:14:31 UTC-08:00
Last update    sent    : 2021/03/15 19:14:31 UTC-08:00
Minimum route advertisement interval is 30 seconds
Optional capabilities:
Route refresh capability has been enabled
4-byte-as capability has been enabled
```

```
Connect-interface has been configured
Multi-hop ebgp has been enabled
Peer Preferred Value: 0
Routing policy configured:
No routing policy is configured
```

(6) 显示 BGP 路由表,以 AR4 为例,结果如下:

```
[AR4]display bgp routing-table
BGP Local router ID is 4.4.4.4
Status codes: * - valid, > - best, d - damped,
              h - history, i - internal, s - suppressed, S - Stale
              Origin : i - IGP, e - EGP, ? - incomplete
Total Number of Routes: 22
```

	Network	NextHop	MED	LocPrf	PrefVal	Path/Ogn
*>	1.1.1.1/32	0.0.0.0	3		0	?
*		5.5.5.5	1	0	200	?
*>	2.2.2.2/32	0.0.0.0	2		0	?
*		5.5.5.5	1	0	200	?
*>	3.3.3.3/32	0.0.0.0	1		0	?
*		5.5.5.5	1	0	200	?
*>	4.4.4.4/32	0.0.0.0	0		0	?
*		5.5.5.5	1	0	200	?
*>	5.5.5.5/32	0.0.0.0	1		0	?
		5.5.5.5	0	0	200	?
*>	6.6.6.6/32	0.0.0.0	1		0	?
*		5.5.5.5	1	0	200	?
*>	192.168.12.0	0.0.0.0	3		0	?
*		5.5.5.5	1	0	200	?
*>	192.168.23.0	0.0.0.0	2		0	?
*		5.5.5.5	1	0	200	?
*>	192.168.34.0	0.0.0.0	0		0	?
*		5.5.5.5	1	0	200	?
*>	192.168.45.0	0.0.0.0	0		0	?
*		5.5.5.5	0	0	200	?
*>	192.168.56.0	0.0.0.0	1		0	?
*		5.5.5.5	0	0	200	?

(7) 验证连通性即测试路由可达,由 AR1 ping AR6,结果如图 4-25 所示。

```
[AR1]ping 192.168.56.6
 PING 192.168.56.6: 56  data bytes, press CTRL_C to break
   Reply from 192.168.56.6: bytes=56 Sequence=1 ttl=251 time=60 ms
   Reply from 192.168.56.6: bytes=56 Sequence=2 ttl=251 time=60 ms
   Reply from 192.168.56.6: bytes=56 Sequence=3 ttl=251 time=50 ms
   Reply from 192.168.56.6: bytes=56 Sequence=4 ttl=251 time=30 ms
   Reply from 192.168.56.6: bytes=56 Sequence=5 ttl=251 time=60 ms

   --- 192.168.56.6 ping statistics ---
   5 packet(s) transmitted
   5 packet(s) received
   0.00% packet loss
   round-trip min/avg/max = 30/52/60 ms

[AR1]
```

图 4-25　测试路由可达（2）

实验七：以太网三层交换机实现 VLAN 间路由配置

1. 实验目的

（1）理解三层交换机的功能。
（2）理解三层交换机中 Vlanif 的含义。
（3）掌握利用三层交换机实现 VLAN 间路由配置的方法。

2. 基础知识

二层交换机是利用 MAC 地址表进行转发操作的，而三层交换机则是一个带有路由功能的二层交换机，是二者的结合。这里的"三层"是指网络层，三层交换机的优势是既能实现三层的路由功能，又能进行二层的高速转发。三层交换技术的出现，解决了企业网划分子网之后，子网之间必须依赖路由器进行通信的问题，多用于企业网内部。

当目的 IP 地址与源 IP 地址不在同一个三层网段时，发送方会向网关请求 ARP 解析，这个网关往往是三层交换机里的一个地址。三层交换模块会查找 ARP 缓存表，将不在同一个三层网段的 IP 地址所对应的 MAC 地址返回发送方，若 ARP 缓存表中没有，则运用其路由功能，找到下一跳的 MAC 地址。找到后，一方面将该 MAC 地址保存，并将其发送给请求方；另一方面将该 MAC 地址发送到二层交换引擎的 MAC 交换表中，这样，以后就可以进行高速的二层转发了。所以，三层交换机有时被描述为"一次路由，多次交换"。

在实际组网中，一个 VLAN 对应一个三层的网段，三层交换机采用 Vlanif（交换虚拟接口）的方式实现 VLAN 间的通信。Vlanif 是指交换机中的虚拟接口，对应一个 VLAN。Vlanif 配置了 IP 地址，并将其作为该 VLAN 对应网段的网关，其作用类似于路由口。

本实验配置命令如表 4-14 所示。

表 4-14　本实验配置命令

命令格式	命令说明
interface vlan 虚拟局域网号	进入 Vlanif 配置模式
display arp	查看交换机的 ARP 缓存信息
display mac-address	查看交换机的交换表

3. 实验流程

实验流程如图 4-26 所示。

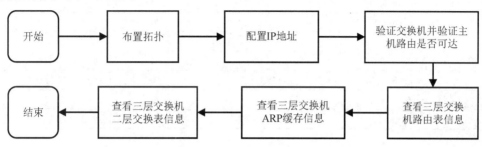

图 4-26　实验流程图

4. 实验步骤

（1）布置拓扑。如图 4-27 所示，网络中共划分了两个三层网段，分别对应两个 VLAN，这种情况下二层交换机是无法配通的，因此可以利用三层交换机使两个网段互通，其 IP 地址配置如表 4-15 所示。

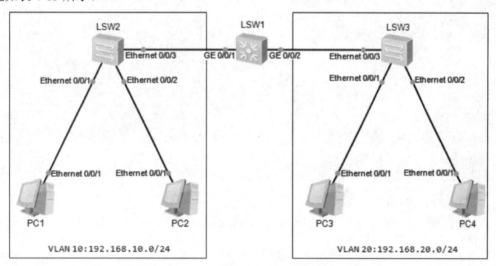

图 4-27　实验拓扑图

表 4-15　各设备的 IP 地址配置

设备名称	接口	IP 地址	VLAN	网关
LSW1	VLAN 10（Vlanif）	192.168.10.254	10	
	VLAN 20（Vlanif）	192.168.20.254	20	
	GE0/0/1		10	
	GE0/0/2		20	
LSW2	Ethernet0/0/1		10	
	Ethernet0/0/2		10	
	Ethernet0/0/3		10	

续表

设备名称	接口	IP 地址	VLAN	网关
LSW3	Ethernet0/0/1		20	
	Ethernet0/0/2		20	
	Ethernet0/0/3		20	
PC1	Ethernet0/0/1	192.168.10.1/24	10	192.168.10.254
PC2	Ethernet0/0/1	192.168.10.2/24	10	192.168.10.254
PC3	Ethernet0/0/1	192.168.20.1/24	20	192.168.20.254
PC4	Ethernet0/0/1	192.168.20.2/24	20	192.168.20.254

（2）配置三层交换机 LSW1。配置如下：

```
[LSW1]vlan batch 10 20    //创建VLAN 10、VLAN 20。
Info: This operation may take a few seconds. Please wait for a moment...done.
[LSW1]int Vlanif 10
//进入VLAN 10接口模式，此接口为虚接口（Vlanif），作为VLAN 10的默认网关。
[LSW1-Vlanif10]ip add 192.168.10.254 24
//给Vlanif配置IP地址。
[LSW1-Vlanif10]q
[LSW1]int Vlanif 20
[LSW1-Vlanif20]ip add 192.168.20.254 24
[LSW1]int g0/0/1
[LSW1-GigabitEthernet0/0/1]port link-type access
[LSW1-GigabitEthernet0/0/1]port default vlan 10
[LSW1-GigabitEthernet0/0/1]int g0/0/2
[LSW1-GigabitEthernet0/0/2]port link-type access
[LSW1-GigabitEthernet0/0/2]port default vlan 20
```

（3）配置 LSW2 和 LSW3 两个二层交换机，以 LSW2 为例：

```
[LSW2]vlan 10
[LSW2]int eth 0/0/1
[LSW2-Ethernet0/0/1]port link-type access
[LSW2-Ethernet0/0/1]port default vlan 10
[LSW2-Ethernet0/0/1]int eth 0/0/2
[LSW2-Ethernet0/0/2]port link-type access
[LSW2-Ethernet0/0/2]port default vlan 10
[LSW2]int eth 0/0/3
[LSW2-Ethernet0/0/3]port link-type access
[LSW2-Ethernet0/0/3]port default vlan 10
```

（4）查看三层交换机路由表信息，结果如下：

```
[LSW1]display ip routing-table
Route Flags: R - relay, D - download to fib
------------------------------------------------------------
Routing Tables: Public
         Destinations : 6        Routes : 6
Destination/Mask    Proto  Pre  Cost  Flags  NextHop         Interface
      127.0.0.0/8   Direct  0    0     D     127.0.0.1       InLoopBack0
      127.0.0.1/32  Direct  0    0     D     127.0.0.1       InLoopBack0
   192.168.10.0/24  Direct  0    0     D     192.168.10.254  Vlanif10
 192.168.10.254/32  Direct  0    0     D     127.0.0.1       Vlanif10
   192.168.20.0/24  Direct  0    0     D     192.168.20.254  Vlanif20
 192.168.20.254/32  Direct  0    0     D     127.0.0.1       Vlanif20
```

通过查看路由表可知，两个 Vlanif 虚接口连接了两个直连网络。需要注意，若不开启三层交换机路由功能，则路由表是空的。

（5）查看三层交换机 ARP 缓存信息。为了便于观察，先将主机互相 ping 通，再来观察 ARP 缓存信息。结果如下：

```
[LSW1]display arp
IP ADDRESS       MAC ADDRESS     EXPIRE(M) TYPE INTERFACE    VPN-INSTANCE   VLAN
------------------------------------------------------------------------------
192.168.10.254   4c1f-cc22-701e            I  - Vlanif10
192.168.10.1     5489-9815-2528     20     D-0  GE0/0/1                     10
192.168.20.254   4c1f-cc22-701e            I  - Vlanif20
192.168.20.1     5489-980c-7c99      5     D-0  GE0/0/2                     20
192.168.20.2     5489-98ce-74c4     20     D-0  GE0/0/2                     20
------------------------------------------------------------------------------
Total:5          Dynamic:3       Static:0       Interface:2
```

可以看到，即便是不同的目的网络，也可以查询到其对应的 MAC 地址，便于进行二层封装，以达到"一次路由，多次交换"的效果。

（6）查看三层交换机二层交换表信息。将 MAC 地址封装为 MAC 帧后，再根据二层交换表将帧转发出去，最终找到目的主机。结果如下：

```
[LSW1]display mac-address
MAC address table of slot 0:
------------------------------------------------------------------------------
MAC Address    VLAN/        PEVLAN CEVLAN Port         Type        LSP/LSR-ID
```

```
VSI/SI                                          MAC-Tunnel
--------------------------------------------------------------------
5489-98ce-74c4  20          -       -       GE0/0/2     dynamic    0/-
5489-9815-2528  10          -       -       GE0/0/1     dynamic    0/-
5489-980c-7c99  20          -       -       GE0/0/2     dynamic    0/-
--------------------------------------------------------------------
Total matching items on slot 0 displayed = 3
```

对于以上过程，读者还可以结合二层交换机的交换表，使用 Wireshark 抓取数据包来仔细观察、分析。

实验八：路由器单臂路由实现 VLAN 间通信

1. 实验目的

（1）理解路由器单臂路由的含义。
（2）掌握路由器单臂路由的配置方法。

2. 基础知识

路由器包含的接口数量一般都比较少，有时为了拓展功能，会将某一个物理接口在逻辑上划分为多个子接口，这些逻辑子接口不能被单独开启或关闭，也就是说，当物理接口被开启或关闭时，所有该接口的子接口也随之被开启或关闭。在实际应用中，往往用这些子接口分别作为局域网中不同 VLAN 的网关，这样就可以仅使用一个物理接口为局域网中不同的 VLAN 提供路由了。这样做的好处是可以节约设备，降低组网成本。

子接口的命名采用如下形式，比如，对于物理接口"GE0/0/1"，其子接口可以命名为"GE0/0/1.1""GE0/0/1.2""GE0/0/1.3"等。

本实验配置命令如表 4-16 所示。

表 4-16 本实验配置命令

命令格式	命令说明
[AR1-GigabitEthernet0/0/0.1]dot1q termination vid 10	给路由器子接口封装 dot1q 协议，并将其划入 VLAN 10

3. 实验流程

实验流程如图 4-28 所示。

4. 实验步骤

（1）布置拓扑。如图 4-29 所示，网络中共划分了三个三层网段，分别对应三个 VLAN，它们都连接在同一个二层交换机上，这里利用路由器单臂路由使三个网段互通，其 IP 地址配置如表 4-17 所示。

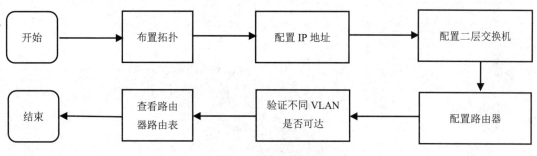

图 4-28 实验流程图

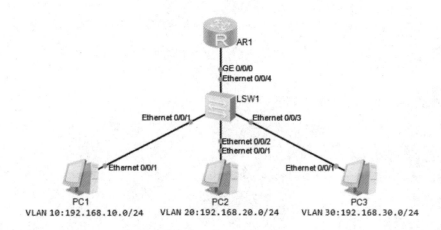

图 4-29 实验拓扑图

表 4-17 各设备的 IP 地址配置

设备名称	接口	IP 地址	VLAN	网关
AR1	GE0/0/0.1	192.168.10.254	10	
	GE0/0/0.2	192.168.20.254	20	
	GE0/0/0.3	192.168.30.254	30	
LSW1	Ethernet0/0/1		10	
	Ethernet0/0/2		20	
	Ethernet0/0/3		30	
	Ethernet0/0/4		Trunk	
PC1	Ethernet0/0/1	192.168.10.1/24	10	192.168.10.254
PC2	Ethernet0/0/1	192.168.20.1/24	20	192.168.20.254
PC3	Ethernet0/0/1	192.168.30.1/24	30	192.168.30.254

（2）配置交换机 LSW1。配置如下：

```
[LSW1]vlan batch 10 20 30
Info: This operation may take a few seconds. Please wait for a moment...done.
[LSW1]int eth 0/0/1
```

```
[LSW1-Ethernet0/0/1]port link-type access
[LSW1-Ethernet0/0/1]port default vlan 10
[LSW1-Ethernet0/0/1]int eth 0/0/2
[LSW1-Ethernet0/0/2]port link-type access
[LSW1-Ethernet0/0/2]port default vlan 20
[LSW1-Ethernet0/0/2]int eth0/0/3
[LSW1-Ethernet0/0/3]port link-type access
[LSW1-Ethernet0/0/3]port default vlan 30
[LSW1-Ethernet0/0/3]int eth 0/0/4
[LSW1-Ethernet0/0/4]port link-type trunk
//因为 VLAN 10、VLAN 20 和 VLAN 30 的流量都通过该接口，所以将其设为 Trunk 模式。
[LSW1-Ethernet0/0/4]port trunk all vlan 10 20 30
[LSW1-Ethernet0/0/4]undo port trunk allow-pass vlan 1
```

(3) 配置 AR1 的路由。配置如下：

```
[AR1]int g0/0/0.1
//进入 g0/0/0 接口的子接口 g0/0/0.1。
[AR1-GigabitEthernet0/0/0.1]dot1q termination vid 10
//给路由器子接口封装 dot1q 协议，并将其划入 VLAN 10。
[AR1-GigabitEthernet0/0/0.1]arp broadcast enable
//开启子接口的 ARP 广播功能。
[AR1-GigabitEthernet0/0/0.1]ip add 192.168.10.254 24
[AR1-GigabitEthernet0/0/0.1]int g0/0/0.2
[AR1-GigabitEthernet0/0/0.2]dot1q termination vid 20
[AR1-GigabitEthernet0/0/0.2]arp broadcast enable
[AR1-GigabitEthernet0/0/0.2]ip add 192.168.20.254 24
[AR1-GigabitEthernet0/0/0.2]int g0/0/0.3
[AR1-GigabitEthernet0/0/0.3]dot1q termination vid 30
[AR1-GigabitEthernet0/0/0.3]arp broadcast enable
[AR1-GigabitEthernet0/0/0.3]ip add 192.168.30.254 24
```

(4) 验证主机路由是否可达。三台主机两两都可 ping 通，请读者自行验证。
(5) 查看路由器路由表，结果如下：

```
[AR1]display ip routing-table
Route Flags: R - relay, D - download to fib
------------------------------------------------------------------------
Routing Tables: Public
         Destinations : 13       Routes : 13
```

```
Destination/Mask       Proto   Pre  Cost  Flags  NextHop         Interface
     127.0.0.0/8       Direct  0    0     D      127.0.0.1       InLoopBack0
     127.0.0.1/32      Direct  0    0     D      127.0.0.1       InLoopBack0
127.255.255.255/32     Direct  0    0     D      127.0.0.1       InLoopBack0
  192.168.10.0/24      Direct  0    0     D      192.168.10.254  GigabitEthernet0/0/0.1
 192.168.10.254/32     Direct  0    0     D      127.0.0.1       GigabitEthernet0/0/0.1
 192.168.10.255/32     Direct  0    0     D      127.0.0.1       GigabitEthernet0/0/0.1
  192.168.20.0/24      Direct  0    0     D      192.168.20.254  GigabitEthernet0/0/0.2
 192.168.20.254/32     Direct  0    0     D      127.0.0.1       GigabitEthernet0/0/0.2
 192.168.20.255/32     Direct  0    0     D      127.0.0.1       GigabitEthernet0/0/0.2
  192.168.30.0/24      Direct  0    0     D      192.168.30.254  GigabitEthernet0/0/0.3
 192.168.30.254/32     Direct  0    0     D      127.0.0.1       GigabitEthernet0/0/0.3
 192.168.30.255/32     Direct  0    0     D      127.0.0.1       GigabitEthernet0/0/0.3
255.255.255.255/32     Direct  0    0     D      127.0.0.1       InLoopBack0
```

通过查看路由表可知，三个网段均属于路由器的直连网段。

实验九：PPP 协议配置（点对点信道）

1. 实验目的

（1）理解 PPP 协议。
（2）掌握不带认证的 PPP 协议配置。
（3）掌握 PAP 和 CHAP 认证的 PPP 协议配置。

2. 基础知识

点对点协议（Point to Point Protocol，PPP）为在点对点连接上传输多协议数据包提供了一个标准方法，是一种点到点的串行通信协议。这种串行链路提供全双工操作，并按照顺序传递数据包。

PPP 协议提供认证的功能，有两种方式，一种是 PAP 认证方式，一种是 CHAP 认证方式。相对来说，PAP 认证方式的安全性没有 CHAP 认证方式高。PAP 认证方式在传输过程中传输的密码（password）是明文的，而 CHAP 认证方式在传输过程中不传输密码，取代密码的是 hash（哈希值）。PAP 认证方式是通过两次握手实现的，而 CHAP 认证方式则是通过 3 次握手实现的。

AAA 是 Authentication（认证）、Authorization（授权）和 Accounting（计费）的简称，是网络安全的一种管理机制，提供了认证、授权、计费 3 种安全功能。这 3 种安全功能的具体作用如下。

（1）认证：验证用户是否可以获得网络访问权。
（2）授权：授权用户可以使用哪些服务。

（3）计费：记录用户使用网络资源的情况。

用户可以使用 AAA 提供的一种或多种安全服务。例如，公司如果仅仅想让员工在访问某些特定资源的时候进行身份认证，那么网络管理员只要配置认证服务器即可；但是如果希望对员工使用网络的情况进行记录，那么还需要配置计费服务器。

如上所述，AAA 是一种管理框架，它提供了授权部分用户去访问特定资源，同时可以记录这些用户操作行为的一种安全机制。由于 AAA 具有良好的可扩展性，并且容易实现用户信息的集中管理，因此被广泛使用。AAA 提供了对用户进行认证、授权和计费等安全功能，防止非法用户登录设备，增强了设备系统的安全性。

更多详细内容请参考教材《计算机网络（第 8 版）》3.2 节。

本实验配置命令如表 4-18 所示。

表 4-18 本实验配置命令

命令格式	命令说明
link-protocol PPP{HDLC}	封装指定协议
ppp authentication chap{ppp}	指定 PPP 用户认证方式
ppp pap local-user 对方路由器名称 password 对方路由器密码	在本路由器上记录对方路由器的名称和密码
aaa	进入 AAA 视图
local-user 用户名 password cipher 密码	创建本地账号，并配置本地账号的登录密码
local-user 用户名 service-type ppp{chap}	配置本地用户使用的服务类型为 PPP{CHAP}

3. 实验流程

实验流程如图 4-30 所示。

图 4-30 实验流程图

4. 实验步骤

（1）布置拓扑。如图 4-31 所示，两台路由器之间用串口相连，若无串口可先关机，等添加了 2SA 串口模块之后再开机。两个串口用 PPP 协议封装，作适当配置使其互通。其 IP 地址配置如表 4-19 所示。

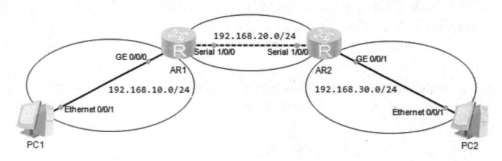

图 4-31 实验拓扑图

表 4-19 各设备的 IP 地址配置

设备名称	接口	IP 地址	网关
PC1	Ethernet0/0/1	192.168.10.1/24	192.168.10.254
AR1	GE0/0/0	192.168.10.254/24	
	Serial1/0/0	192.168.20.1/24	
AR2	Serial1/0/0	192.168.20.2/24	
	GE0/0/0	192.168.30.254/24	
PC2	Ethernet0/0/1	192.168.30.1/24	192.168.30.254

(2) 配置路由。

配置 AR1 的路由。配置如下：

```
[Huawei]sy AR1
[AR1]int g0/0/0
[AR1-GigabitEthernet0/0/0]ip add 192.168.10.254 24
[AR1-GigabitEthernet0/0/0]ints1/0/0
[AR1-Serial1/0/0]ip address 192.168.20.1 24
[AR1]ospf 1 router-id 1.1.1.1
[AR1-ospf-1]area 0
[AR1-ospf-1-area-0.0.0.0]network 192.168.20.1 0.0.0.0
[AR1-ospf-1-area-0.0.0.0]network 192.168.10.254 0.0.0.0
```

配置 AR2 的路由。配置如下：

```
[Huawei]sy AR2
[AR2]int s 1/0/0
[AR2-Serial1/0/0]
[AR2-Serial1/0/0]ip add 192.168.20.2 24
[AR2-Serial1/0/0]int g 0/0/1
[AR2-GigabitEthernet0/0/1]ip add 192.168.30.254 24
[AR2-GigabitEthernet0/0/1]q
[AR2]ospf 1 router-id 2.2.2.2
[AR2-ospf-1]area 0
[AR2-ospf-1-area-0.0.0.0]network 192.168.30.254 0.0.0.0
[AR2-ospf-1-area-0.0.0.0]network 192.168.20.2 0.0.0.0
```

经过路由配置后，两台 PC 是可以 ping 通的，并且两台路由器之间的串行链路封装了 PPP 协议，这是因为华为路由器串行接口默认封装了 PPP 协议。查看串行接口的信息，结果如下：

```
[AR1]display interface Serial 1/0/0
Serial1/0/0 current state : UP
Line protocol current state : UP
```

```
Last line protocol up time : 2020-10-21 20:57:33 UTC-08:00
Description:HUAWEI, AR Series, Serial1/0/0 Interface
Route Port,The Maximum Transmit Unit is 1500, Hold timer is 10(sec)
Internet Address is 192.168.20.1/24
Link layer protocol is PPP
LCP opened, IPCP opened
Last physical up time   : 2020-10-21 20:54:45 UTC-08:00
Last physical down time : 2020-10-21 20:54:45 UTC-08:00
Current system time: 2020-10-21 20:58:58-08:00
Physical layer is synchronous, Virtualbaudrate is 64000 bps
Interface is DTE, Cable type is V11, Clock mode is TC
Last 300 seconds input rate 11 bytes/sec 88 bits/sec 0 packets/sec
Last 300 seconds output rate 7 bytes/sec 56 bits/sec 0 packets/sec

Input: 126 packets, 5060 bytes
  Broadcast:            0,  Multicast:                0
  Errors:               0,  Runts:                    0
  Giants:               0,  CRC:                      0

  Alignments:           0,  Overruns:                 0
  Dribbles:             0,  Aborts:                   0
  No Buffers:           0,  Frame Error:              0

Output: 132 packets, 2836 bytes
  Total Error:          0,  Overruns:                 0
  Collisions:           0,  Deferred:                 0
  Input bandwidth utilization   :    0%
  Output bandwidth utilization  :    0%
```

（3）封装不带认证的 PPP 协议。

实际上，尽管 HDLC 协议也是 ISO 定义的标准，但该标准被不同的厂家进行了扩展，兼容性并不好。

配置 AR1。配置如下：

```
[AR1]int s 1/0/0
[AR1-Serial1/0/0]link-protocol ppp
```

配置 AR2。配置如下：

```
[AR2]int s 1/0/0
```

```
[AR2-Serial1/0/0]link-protocol ppp
```

此时两台 PC 可以 ping 通,并且接口信息已经被封装为 PPP 协议,请读者自行验证。
(4) 封装带 PAP 认证的 PPP 协议。
在路由配通的基础上,做如下配置。
配置 AR1。配置如下:

```
[AR1]int s 1/0/0
[AR1-Serial1/0/0]ppp authentication-mode pap domain huawei
//进行 PAP 认证。
[AR1-Serial1/0/0]ppp pap local-user huawei password 123456
//该用户名和密码为对方 PAP 认证发送的用户名和密码。
[AR1]aaa
[AR1-aaa]local-user huawei password cipher 123456
[AR1-aaa]local-user huawei service-type ppp
//发送认证所需的用户名和密码。
```

配置 AR2。配置如下:

```
[AR2]int s 1/0/0
[AR2-Serial1/0/0]ppp authentication-mode pap domain huawei
//进行 PAP 认证。
[AR2-Serial1/0/0]ppp pap local-user huawei password 123456
//该用户名和密码为对方 PAP 认证发送的用户名和密码。
[AR2]aaa
[AR2-aaa]local-user huawei password cipher 123456
[AR2-aaa]local-user huawei service-type ppp
//发送认证所需的用户名和密码。
```

经验证,两台 PC 可以 ping 通。
(5) 封装带 CHAP 认证的 PPP 协议。
在路由配通的基础上,做如下配置。
配置 AR1。配置如下:

```
[AR1]int s 1/0/0
[AR1-Serial1/0/0]ppp authentication-mode chap
//设置 PPP 认证方式为 CHAP。
[AR1-Serial1/0/0]ppp chap user Huawei
[AR1]aaa
[AR1-aaa]local-user huawei password cipher 123456
[AR1-aaa]local-user huawei service-type ppp
```

配置 AR2。配置如下：

```
[AR2]int s 1/0/0
[AR2-Serial1/0/0]ppp authentication-mode chap
//设置 PPP 认证方式为 CHAP。
[AR2-Serial1/0/0]ppp chap user Huawei
[AR2]aaa
[AR2-aaa]local-user huawei password cipher 123456
[AR2-aaa]local-user huawei service-type ppp
```

需要注意，串口双方的密码都要一致。

此时两台 PC 可以 ping 通，请读者自行验证并查看接口信息。

实验十：访问控制列表（ACL）实验

1. 实验目的

（1）理解 ACL 的含义。
（2）初步掌握 ACL 的配置和应用。

2. 基础知识

访问控制列表（Access Control List，ACL）也称接入控制列表，俗称防火墙，在有的文档中还称包过滤。ACL 通过定义一些规则对网络设备接口上的数据包进行控制。每个 ACL 可以包含多条规则，这些规则被组织在一起成为一个整体。

对 ACL 的命名有两种方式，分别为编号 ACL 和命名 ACL，这些编号或命名是唯一的，在设备配置中将通过引用它们来达到控制访问的目的。

ACL 有两种访问列表，分别为基本 ACL 和高级 ACL，基本 ACL 的编号范围为 2000～2999，高级 ACL 的编号范围为 3000～3999。

基本 ACL 仅使用 IPv4 报文的源 IP 地址、分片信息和生效时间段信息来定义规则。高级 ACL 既可以使用 IPv4 报文的源 IP 地址，也可以使用目的 IP 地址、IP 协议类型、ICMP 类型、TCP 源/目的接口号、UDP 源/目的接口号、生效时间段等信息来定义规则。

不管是基本 ACL 还是高级 ACL，都需要注意以下规则。

（1）在表达源 IP 地址和目的 IP 地址时，经常使用 any。

any 表示允许所有 IP 地址作为源 IP 地址。例如，下面两行是等价的：

```
rule permit 0.0.0.0 255.255.255.255
rule permit any
```

（2）对于有多条规则的 ACL，这些规则的顺序很重要，ACL 严格按生效的顺序进行匹配。可以使用"display acl all"或"display acl 编号"命令查看生效的 ACL 规则顺序。若分组与某条规则相匹配，则根据规则中的关键字 permit 或 deny 进行操作，所有的后续规则均被忽略。也就是说，采用的是首先匹配的算法，路由器从第一条规则开始依次检查列表，一次一条规则，直至发现匹配项。

（3）每个 ACL 的最后，系统都会自动附加一条隐式 deny 的规则，这条规则拒绝所有流量。对于与用户指定的任何规则都不匹配的分组，隐式拒绝规则起到了截流的作用，所有分组均与该规则相匹配。

配置 ACL 需要实施如下步骤。

（1）创建 ACL。

（2）使用 traffic-filter 匹配 ACL。

在应用 ACL 时，还需要指出是进站方向还是出站方向。进站指从外面进入接口时进行检查，出站则相反。

本实验配置命令如表 4-20 所示。

表 4-20 本实验配置命令

命令格式	命令说明
acl 列表号	全局配置模式下定义 ACL
rule permit(deny) source source-ip wildcard-mask	ACL 配置模式下，定义 ACL 规则。其中，permit(deny)为允许(拒绝)，source-ip 和 wildcard-mask 分别为源 IP 地址和通配符掩码
traffic-filter inbound(outbound) acl 列表号	在接口配置模式下，将 ACL 应用到该接口上。inbound 表示数据包从该接口进来时进行检查，outbound 表示数据包从该接口出去时进行检查

3. 实验流程

实验流程如图 4-32 所示。

图 4-32 实验流程图

4. 实验步骤

（1）布置拓扑。如图 4-33 所示，Company 是公司出口路由器，对内用单臂路由连接了三个子网。其中，VLAN 30 里有一台公司内部服务器，只允许公司内部访问；VLAN 10 是内部员工子网，由于工作需要，不允许访问外部网络；VLAN 20 是管理人员子网，允许访问外部网络。这里暂不考虑公有地址和私有地址的差别。其 IP 地址配置如表 4-21 所示。

表 4-21 各设备的 IP 地址配置

设备名称	接口	IP 地址	默认网关
路由器 Company	GE0/0/0.1	172.16.1.254/24	
	GE0/0/0.2	172.16.2.254/24	
	GE0/0/0.3	172.16.3.254/24	
	GE0/0/1	192.168.1.1/24	

续表

设备名称	接口	IP 地址	默认网关
路由器 Internet	GE0/0/0	192.168.2.254/24	
	GE0/0/1	192.168.1.2/24	
PC1	Ethernet0/0/1	172.16.1.1/24	172.16.1.254/24
PC2	Ethernet0/0/1	172.16.2.1/24	172.16.2.254/24
Server1	Ethernet0/0/0	172.16.3.1/24	172.16.3.254/24
Server2	Ethernet0/0/0	192.168.2.1/24	192.168.2.254/24
Server3	Ethernet0/0/0	192.168.2.2/24	192.168.2.254/24

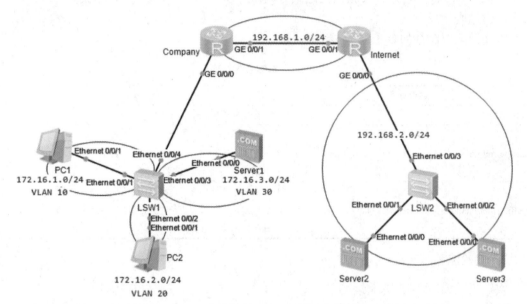

图 4-33　实验拓扑图

（2）配置路由。这里使用 OSPF 配置路由。

配置交换机 LSW1。配置如下：

```
[Huawei]sy LSW1
[LSW1]vlan batch 10 20 30
Info: This operation may take a few seconds. Please wait for a moment...done.
[LSW1]int eth 0/0/1
[LSW1-Ethernet0/0/1]port link-type access
[LSW1-Ethernet0/0/1]port default vlan 10
[LSW1-Ethernet0/0/1]int eth 0/0/2
[LSW1-Ethernet0/0/2]port link-type access
[LSW1-Ethernet0/0/2]port default vlan 20
[LSW1-Ethernet0/0/2]int eth 0/0/3
[LSW1-Ethernet0/0/3]port link-type access
[LSW1-Ethernet0/0/3]port default vlan 30
```

```
[LSW1-Ethernet0/0/3]int eth 0/0/4
[LSW1-Ethernet0/0/4]port link-type trunk
[LSW1-Ethernet0/0/4]port trunk all vlan 10 20 30
[LSW1-Ethernet0/0/4]undo port trunk all vlan 1
```

配置路由器 Company 的路由。配置如下：

```
[Huawei]sy Company
[Company]int g0/0/0.1
[Company-GigabitEthernet0/0/0.1]ip add 172.16.1.254 24
[Company-GigabitEthernet0/0/0.1]dot1q termination vid 10
[Company-GigabitEthernet0/0/0.1]arp broadcast enable
[Company-GigabitEthernet0/0/0.1]int g0/0/0.2
[Company-GigabitEthernet0/0/0.2]ip add 172.16.2.254 24
[Company-GigabitEthernet0/0/0.2]dot1q termination vid 20
[Company-GigabitEthernet0/0/0.2]arp broadcast enable
[Company-GigabitEthernet0/0/0.2]int g0/0/0.3
[Company-GigabitEthernet0/0/0.3]ip add 172.16.3.254 24
[Company-GigabitEthernet0/0/0.3]dot1q termination vid 30
[Company-GigabitEthernet0/0/0.3]arp broadcast enable
[Company-GigabitEthernet0/0/0.3]q
[Company]ospf 1 router-id 1.1.1.1
[Company-ospf-1]area 0
[Company-ospf-1-area-0.0.0.0]network 172.16.1.254 0.0.0.0
[Company-ospf-1-area-0.0.0.0]network 172.16.2.254 0.0.0.0
[Company-ospf-1-area-0.0.0.0]network 172.16.3.254 0.0.0.0
[Company-ospf-1-area-0.0.0.0]network 192.168.1.1 0.0.0.0
```

配置路由器 Internet 的路由。配置如下：

```
[Huawei]sy Internet
[Internet]int g0/0/1
[Internet-GigabitEthernet0/0/1]ip add 192.168.1.2 24
[Internet-GigabitEthernet0/0/1]int g0/0/0
[Internet-GigabitEthernet0/0/0]ip add 192.168.2.254 24
[Internet-GigabitEthernet0/0/0]q
[Internet]ospf 1 router-id 2.2.2.2
[Internet-ospf-1]area 0
[Internet-ospf-1-area-0.0.0.0]network 192.168.1.2 0.0.0.0
[Internet-ospf-1-area-0.0.0.0]network 192.168.2.254 0.0.0.0
```

经过以上配置后，路由两两可达，请读者自行验证。

（3）配置 ACL 并验证效果。为满足要求，此处设计两个 ACL，分别应用在两个接口上。配置路由器 Company 的 ACL。配置如下：

```
[Company]acl 2000
[Company-acl-basic-2000]rule 5 permit source 172.16.1.0 0.0.0.255
[Company-acl-basic-2000]rule 10 permit source 172.16.2.0 0.0.0.255
//创建 acl 2000，允许两个子网通过，但在规则最后自动附加一条隐式 deny 的规则，这条规则拒
//绝所有流量。也就是说，只允许这两个子网访问服务器。
[Company]acl 2001
[Company-acl-basic-2001]rule 5 permit source 172.16.2.0 0.0.0.255
[Company-acl-basic-2001]rule 10 deny
[Company]interface GigabitEthernet0/0/0
[Company-GigabitEthernet0/0/0.3]traffic-filter outbound acl 2000
//将 acl 2000 应用在该接口，出站时检查。
[Company]interface GigabitEthernet0/0/1
[Company-GigabitEthernet0/0/1]traffic-filter outbound acl 2001
//将 acl 2001 应用在该接口，出站时检查。员工子网 VLAN 10 将匹配后面的 deny 规则，从而无
//法从该接口出去。
```

如图 4-34 所示，在没有应用 ACL 之前，员工子网可以访问外部网络的 IP 地址。

```
PC>ping 192.168.2.1

Ping 192.168.2.1: 32 data bytes, Press Ctrl_C to break
From 192.168.2.1: bytes=32 seq=1 ttl=253 time=47 ms
From 192.168.2.1: bytes=32 seq=2 ttl=253 time=47 ms
From 192.168.2.1: bytes=32 seq=3 ttl=253 time=63 ms
From 192.168.2.1: bytes=32 seq=4 ttl=253 time=62 ms
From 192.168.2.1: bytes=32 seq=5 ttl=253 time=63 ms

--- 192.168.2.1 ping statistics ---
  5 packet(s) transmitted
  5 packet(s) received
  0.00% packet loss
  round-trip min/avg/max = 47/56/63 ms
```

图 4-34　ACL 应用前

应用 ACL 之后，显示响应时间超时，如图 4-35 所示。

```
PC>ping 192.168.2.1

Ping 192.168.2.1: 32 data bytes, Press Ctrl_C to break
Request timeout!
Request timeout!
Request timeout!
Request timeout!
Request timeout!

--- 192.168.2.1 ping statistics ---
  5 packet(s) transmitted
  0 packet(s) received
  100.00% packet loss

PC>
```

图 4-35　ACL 应用后

从外部访问公司内部服务器，在 ACL 的作用下，无法 ping 通，如图 4-36 所示。

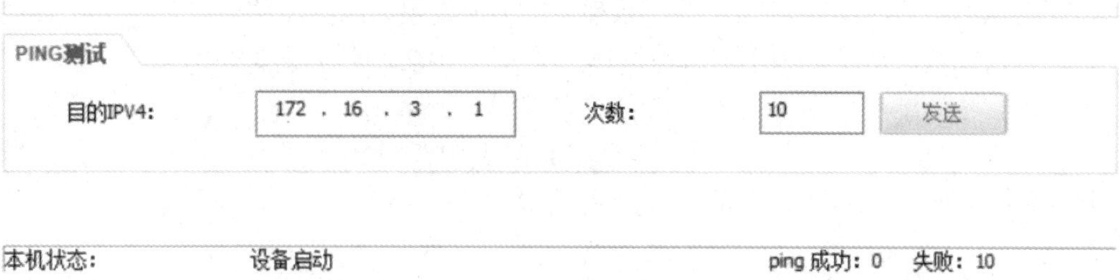

```
PING测试
    目的IPV4:    172 . 16 . 3 . 1    次数:    10    发送

本机状态:      设备启动                    ping 成功: 0   失败: 10
```

图 4-36　外部访问内部服务器

实验十一：网络地址转换（NAT）实验

1. 实验目的

（1）理解 NAT 的含义。
（2）理解 NAT 的三种转换方式。
（3）初步掌握 NAT 的配置和应用。

2. 基础知识

目前，很多局域网内部使用的都是专用地址，这主要是由全球 IP 地址的紧缺造成的。而互联网上的路由器对于目的地址为专用地址的 IP 数据报一律不进行转发，这种情况下，局域网连通互联网主要是采用了 NAT 技术。

网络地址转换（Network Address Translation，NAT）是 1994 年提出的，主要用来解决专用地址和全球地址转换的问题。局域网内部的通信只需要专用地址即可，当访问外部互联网时，专用地址可以转换为一个全球地址（有时也称公网地址）去访问，这种方法需要在局域网连接到互联网的路由器上安装 NAT 软件。装有 NAT 软件的路由器称为 NAT 路由器，它至少有一个有效的外部全球 IP 地址。

使用 NAT 技术的优点如下。

（1）节省公网地址。NAT 可以让局域网内部使用专用地址的主机共用少量的公网地址来访问互联网，而不需要为每一台主机都申请一个 IP 地址，这就在一定程度上节约了公网地址。

（2）由于在访问互联网时专用地址被转换为一个公网地址，因此这就对外部网络屏蔽了内部的网络拓扑，提高了安全性。

由于要经过地址转换的环节，因此采用 NAT 技术会轻微影响网络速度，尽管如此，NAT 技术仍然得到了较广泛的应用。

NAT 包括以下三种技术类型。

（1）静态 NAT：把内部网络中的每个主机地址永久映射成外部网络中的某个合法地址。如果内部网络有对外提供服务的需求，如 WWW 服务器、FTP 服务器等，那么这些服务器的

IP 地址应该采用静态 NAT，以便外部用户可以使用这些服务。静态 NAT 不能节省 IP 地址。

（2）动态 NAT：把外部网络中的一系列公网地址使用动态分配的方法映射到内部网络。转换时，从内部合法地址范围中动态地选择一个未使用的地址与内部专用地址进行转换。当然，当内部合法地址使用完毕时，后续的 NAT 申请将失败。

（3）接口映射：把内部地址映射到一个内部合法 IP 地址的不同接口上，这也是一种动态的地址转换，适用于只申请到少量 IP 地址的情况。

配置动态 IP 地址转换，可参考以下步骤。

（1）配置路由，确保路由可达。

（2）设计标准的 IP 访问控制列表，规定哪些 IP 地址可以被转换。

（3）设计 NAT 地址池，规定可被转换的公网地址。

（4）将 ACL 映射到 NAT 地址池。

（5）启用 NAT。

更多详细内容请参考教材《计算机网络（第 8 版）》4.8.2 节。

本实验配置命令如表 4-22 所示。

表 4-22 本实验配置命令

命令格式	命令说明
nat static global X.X.X.X inside X.X.X.X	静态 NAT。第一个地址为转换后的地址，第二个地址为转换前的地址
ip pool pool-name	创建全局地址池，pool-name 为地址池的名字
nat address-group X X.X.X.X X.X.X.X	配置 NAT 地址池
acl X	创建一个 ACL
rule permit source X.X.X.X X	配置基本 ACL 的规则
nat outbound XXX address-group X	用来将一个 ACL 和一个 NAT 地址池关联起来，表示 ACL 中规定的地址可以使用 NAT 地址池进行地址转换

3. 实验流程

实验流程如图 4-37 所示。

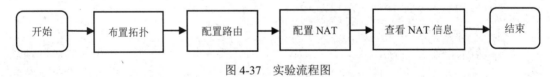

图 4-37 实验流程图

4. 静态 NAT 实验步骤

（1）布置拓扑。如图 4-38 所示，公司内部有两台 PC，其中，PC1 访问外网时其 IP 地址将被转换成 100.100.100.1，PC2 访问外网时其 IP 地址将被转换成 100.100.100.2，AR1 是公司出口路由器，NAT 被部署在这里，专用地址通往 ISP 时在这里被转换为公网地址，交换机 LSW1 不需要配置。其 IP 地址配置如表 4-23 所示。

第 4 章 网络层

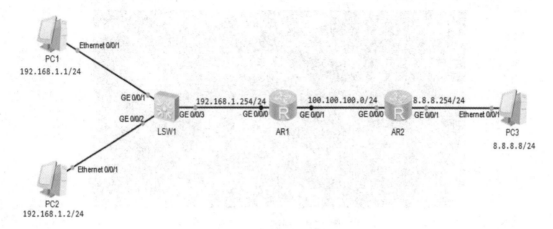

图 4-38 实验拓扑图（1）

表 4-23 各设备的 IP 地址配置（1）

设备名称	接口	IP 地址	默认网关
AR1	GE0/0/0	192.168.1.254/24	
	GE0/0/1	100.100.100.253/24	
AR2	GE0/0/0	100.100.100.254/24	
	GE0/0/1	8.8.8.254/24	
PC1		192.168.1.1/24	192.168.1.254/24
PC2		192.168.1.2/24	192.168.1.254/24
PC3		8.8.8.8/24	8.8.8.254/24

（2）配置路由。

配置 AR1 的路由。配置如下：

```
[Huawei]sy AR1
[AR1]int g0/0/0
[AR1-GigabitEthernet0/0/0]ip add 192.168.1.254 24
[AR1-GigabitEthernet0/0/0]int g0/0/1
[AR1-GigabitEthernet0/0/1]ip add 100.100.100.253 24
[AR1]ip route-static 0.0.0.0 0.0.0.0 100.100.100.254
```

配置 AR2 的路由。配置如下：

```
[Huawei]sy AR2
[AR2]int g0/0/0
[AR2-GigabitEthernet0/0/0]ip add 100.100.100.254 24
[AR2-GigabitEthernet0/0/0]int g0/0/1
[AR2-GigabitEthernet0/0/1]ip add 8.8.8.254 24
```

经过以上配置后，网络可以全部 ping 通。由 PC1 ping PC3，其结果如图 4-39 所示。

```
PC>ping 8.8.8.8
Ping 8.8.8.8: 32 data bytes, Press Ctrl_C to break
From 8.8.8.8: bytes=32 seq=1 ttl=126 time=47 ms
From 8.8.8.8: bytes=32 seq=2 ttl=126 time=47 ms
From 8.8.8.8: bytes=32 seq=3 ttl=126 time=31 ms
From 8.8.8.8: bytes=32 seq=4 ttl=126 time=47 ms
From 8.8.8.8: bytes=32 seq=5 ttl=126 time=47 ms

--- 8.8.8.8 ping statistics ---
  5 packet(s) transmitted
  5 packet(s) received
  0.00% packet loss
  round-trip min/avg/max = 31/43/47 ms
```

图 4-39　测试结果（1）

（3）在 AR1 上配置 NAT。配置如下：

```
[AR1-GigabitEthernet0/0/1]nat static global 100.100.100.1 inside 192.168.1.1
[AR1-GigabitEthernet0/0/1]nat static global 100.100.100.2 inside 192.168.1.2
//为 PC 定义一个静态转换的公网地址。
```

（4）查看 NAT 转换记录。

首先使用 Wireshark 对 AR1 的 GE0/0/1 接口进行抓包，然后由 PC1 ping PC3，查看转换记录，结果如下：

```
No.    Time        Source           Destination        Protocol  Info
  7  6.000000    100.100.100.1      8.8.8.8            ICMP      Echo (ping)
request  (id=0x5e4a, seq(be/le)=4/1024, ttl=127)

Frame 7: 74 bytes on wire (592 bits), 74 bytes captured (592 bits)
  Ethernet  II,  Src:  HuaweiTe_0d:53:a3  (00:e0:fc:0d:53:a3),  Dst:
HuaweiTe_a2:34:57 (00:e0:fc:a2:34:57)
  Internet Protocol, Src: 100.100.100.1 (100.100.100.1), Dst: 8.8.8.8 (8.8.8.8)
  Internet Control Message Protocol


No.    Time        Source           Destination        Protocol  Info
  8  6.016000    8.8.8.8            100.100.100.1      ICMP      Echo (ping)
reply    (id=0x5e4a, seq(be/le)=4/1024, ttl=127)

Frame 8: 74 bytes on wire (592 bits), 74 bytes captured (592 bits)
  Ethernet  II,  Src:  HuaweiTe_a2:34:57  (00:e0:fc:a2:34:57),  Dst:
HuaweiTe_0d:53:a3 (00:e0:fc:0d:53:a3)
  Internet Protocol, Src: 8.8.8.8 (8.8.8.8), Dst: 100.100.100.1 (100.100.100.1)
  Internet Control Message Protocol
```

5. 动态 NAT 实验步骤

（1）布置拓扑。如图 4-40 所示，公司内部有两台 PC，其中 PC1 访问外网时，其 IP 地址将被动态转换成 100.100.100.3～100.100.100.10 中的一个，PC2 不允许访问外网，AR1 是公司出口路由器，NAT 被部署在这里，专用地址通往 ISP 时在这里被转换为公网地址，交换机 LSW1 不需配置。其 IP 地址配置如表 4-24 所示。

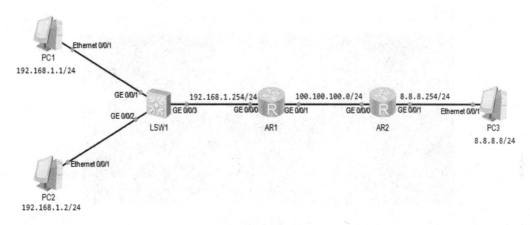

图 4-40　实验拓扑图（2）

表 4-24　各设备的 IP 地址配置（2）

设备名称	接口	IP 地址	默认网关
AR1	GE0/0/0	192.168.1.254/24	
	GE0/0/1	100.100.100.1/24	
AR2	GE0/0/0	100.100.100.2/24	
	GE0/0/1	8.8.8.254/24	
PC1		192.168.1.1/24	192.168.1.254/24
PC2		192.168.1.2/24	192.168.1.254/24
PC3		8.8.8.8/24	8.8.8.254/24

（2）配置路由。

配置 AR1 的路由。配置如下：

```
[Huawei]sy AR1
[AR1]int g0/0/0
[AR1-GigabitEthernet0/0/0]ip add 192.168.1.254 24
[AR1-GigabitEthernet0/0/0]int g0/0/1
[AR1-GigabitEthernet0/0/1]ip add 100.100.100.1 24
[AR1-GigabitEthernet0/0/1]q
[AR1]ip route-static 0.0.0.0 0.0.0.0 100.100.100.2
```

配置 AR2 的路由。配置如下：

```
[Huawei]sy AR2
[AR2]int g0/0/1
[AR2-GigabitEthernet0/0/1]ip add 8.8.8.254 24
[AR2-GigabitEthernet0/0/1]int g0/0/0
[AR2-GigabitEthernet0/0/0]ip add 100.100.100.2 24
```

经过以上配置后，网络可以全部 ping 通。由 PC1 ping PC3，其结果如图 4-41 所示。

```
PC>ping 8.8.8.8

Ping 8.8.8.8: 32 data bytes, Press Ctrl_C to break
From 8.8.8.8: bytes=32 seq=1 ttl=126 time=47 ms
From 8.8.8.8: bytes=32 seq=2 ttl=126 time=31 ms
From 8.8.8.8: bytes=32 seq=3 ttl=126 time=47 ms
From 8.8.8.8: bytes=32 seq=4 ttl=126 time=47 ms
From 8.8.8.8: bytes=32 seq=5 ttl=126 time=31 ms

--- 8.8.8.8 ping statistics ---
  5 packet(s) transmitted
  5 packet(s) received
  0.00% packet loss
  round-trip min/avg/max = 31/40/47 ms
```

图 4-41 测试结果（2）

（3）在 AR1 上配置 NAT。配置如下：

```
[AR1]ip pool jiaoxue
Info: It's successful to create an IP address pool.
[AR1-ip-pool-jiaoxue]q
[AR1]nat address-group 1 100.100.100.3 100.100.100.10
//定义公网地址池。
[AR1]acl 2000
[AR1-acl-basic-2000]rule permit source 192.168.1.1 0
//配置 ACL，定义可被转换的专用地址。
[AR1-acl-basic-2000]q
[AR1]int g0/0/1
[AR1-GigabitEthernet0/0/1]nat outbound 2000 address-group 1
//将 NAT 地址池与 ACL 关联起来。
```

（4）查看 NAT 转换记录。

首先使用 Wireshark 对 AR1 的 GE0/0/1 接口进行抓包，然后由 PC1 ping PC3，查看转换记录，结果如下：

```
No.    Time       Source            Destination       Protocol    Info
2      2.000000   100.100.100.3     8.8.8.8           ICMP        Echo (ping)
request (id=0x1028, seq(be/le)=2/512, ttl=127)
```

Frame 2: 74 bytes on wire (592 bits), 74 bytes captured (592 bits)
Ethernet II, Src: HuaweiTe_0a:4f:10 (00:e0:fc:0a:4f:10), Dst: HuaweiTe_a2:5a:8c (00:e0:fc:a2:5a:8c)
Internet Protocol, **Src: 100.100.100.3 (100.100.100.3), Dst: 8.8.8.8 (8.8.8.8)**
Internet Control Message Protocol

No.	Time	Source	Destination	Protocol	Info
3	2.000000	**8.8.8.8**	**100.100.100.3**	ICMP	Echo (ping) reply (id=0x1028, seq(be/le)=2/512, ttl=127)

Frame 3: 74 bytes on wire (592 bits), 74 bytes captured (592 bits)
Ethernet II, Src: HuaweiTe_a2:5a:8c (00:e0:fc:a2:5a:8c), Dst: HuaweiTe_0a:4f:10 (00:e0:fc:0a:4f:10)
Internet Protocol, **Src: 8.8.8.8 (8.8.8.8), Dst: 100.100.100.3 (100.100.100.3)**
Internet Control Message Protocol

第 5 章 运输层

实验一：TCP 连接实验

1. 实验目的

理解 TCP 连接过程。

2. 基础知识

传输控制协议（Transmission Control Protocol，TCP）是运输层的两个主要协议之一，是面向连接的协议，即双方在通信之前必须要先建立连接，通信结束后必须要释放连接。

TCP 在建立连接的过程中，客服双方要交换三个报文段，这就是所谓的"三报文握手"。采用三报文握手的原因：客户端发出的连接请求报文可能会延迟到达服务器，在这段时间里，客户端会因超时等因素重新发出新的连接请求；而对服务器来说，就有可能会收到两个连接请求，且其中一个显然是失效的，不应该建立连接。如果采用二报文握手的机制，那么就会建立两个连接，这样就消耗了服务器的资源。

三报文握手过程如图 5-1 所示。

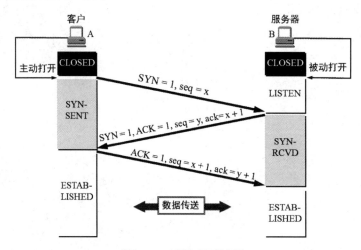

图 5-1 三报文握手过程

首先，客户端向服务器的 TCP 进程发出连接请求报文段，这时首部的同步位 SYN=1，同时选择一个初始序号 seq=x，客户端状态为 SYN-SENT。

然后，服务器收到连接请求报文之后，若同意建立连接，则向客户端发送确认报文。在确认报文段中 SYN 位和 ACK 位都置 1，确认号 ack=x+1，同时初始序号 seq=y。

最后，客户端收到服务器的确认后，还要向服务器给出确认，确认报文段的 ACK 位置 1，ack=y+1，自己的序号 seq=x+1。

详细内容请参考教材《计算机网络（第 8 版）》5.9.1 节。

3. 实验流程

实验流程如图 5-2 所示。

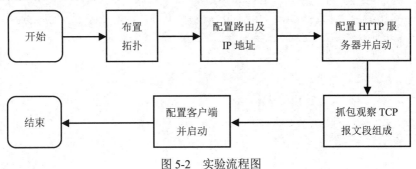

图 5-2　实验流程图

4. 实验步骤

（1）实验拓扑如图 5-3 所示。

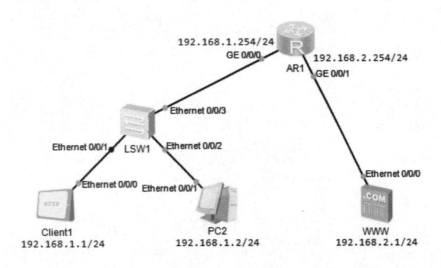

图 5-3　实验拓扑图

（2）配置 IP 地址及路由，确保 Client1 能 ping 通。

（3）使用 Wireshark 抓取数据包，并观察、分析 TCP 协议。首先，打开 WWW 服务器的页面，单击服务器信息，单击 HttpServer，选择文件根目录（桌面即可）；然后，右击 Client1 的 Ethernet0/0/0 接口，单击"开始抓包"；最后，打开 Client1 客户端的页面，单击客户端信息，单击 HttpServer 并输入 WWW 服务器的 IP 地址。因为应用层的 HTTP 在运输层使用 TCP 协议，所以在 Client1 处封装了 TCP 报文段。由于双方需要建立 TCP 连接，因此可通过双击 TCP 报文段来观察 TCP 报文段的内容，体会三报文握手的过程。过程如下：

```
No.     Time         Source          Destination      Protocol Info
1       0.000000     192.168.1.1     192.168.2.1      TCP
```

```
lot105-ds-upd > http [SYN] Seq=0 Win=8192 Len=0 MSS=1460
```
//上述结果为第一次报文握手封装的TCP报文段，其SYN=1，seq=0。

No.	Time	Source	Destination	Protocol Info
2	0.047000	192.168.2.1	192.168.1.1	TCP

```
http > lot105-ds-upd [SYN, ACK] Seq=0 Ack=1 Win=8192 Len=0 MSS=1460
```
//上述结果为WWW服务器的出站TCP报文段，属于第二次报文握手，其SYN=1，ACK=1，seq=0，
//ack=1。

No.	Time	Source	Destination	Protocol Info
3	0.047000	192.168.1.1	192.168.2.1	TCP

```
lot105-ds-upd > http [ACK] Seq=1 Ack=1 Win=8192 Len=0
```
//上述结果为PC封装的TCP报文段，属于第三次报文握手，其ACK=1，seq=0+1=1，ack=1。

经过三报文握手后，开始传输数据。

第 6 章　应用层

实验一：远程终端协议（TELNET）实验

1. 实验目的

（1）理解 TELNET 的含义。
（2）掌握利用 TELNET 登录路由器的方法。

2. 基础知识

TELNET 协议是 TCP/IP 协议簇中的一员，是 Internet 远程登录服务的标准协议和主要方式。它可以在本地计算机上登录并操控远程主机，对用户来说，就好像直接在操控远程主机一样。

更多详细内容请参考教材《计算机网络（第 8 版）》6.3 节。

3. 实验流程

实验流程如图 6-1 所示。在本实验中，由路由器 AR1 远程登录路由器 AR3，之后可在 AR1 上对路由器进行操作配置（华为模拟器 PC 暂时不支持 TELNET）。

图 6-1　实验流程图

4. 实验步骤

（1）按图 6-2 所示布置拓扑，为了简单明确，网络中设了两个网段，分别为 192.168.1.0/24 和 192.168.2.0/24。这个实验的关键是确保路由可达。

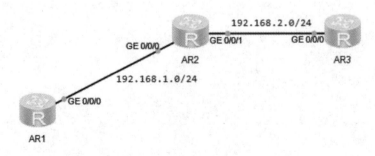

图 6-2　实验拓扑图

（2）配置路由。

配置 AR1 的路由。配置如下：

```
[Huawei] sy AR1
[AR1]int g0/0/0
[AR1-GigabitEthernet0/0/0]ip add 192.168.1.1 24
[AR1-GigabitEthernet0/0/0]q
[AR1]ospf 1 router-id 1.1.1.1
[AR1-ospf-1]area 0
[AR1-ospf-1-area-0.0.0.0]network 192.168.1.1 0.0.0.0
```

配置 AR2 的路由。配置如下：

```
[Huawei]sy AR2
[AR2]int g0/0/0
[AR2-GigabitEthernet0/0/0]ip add 192.168.1.2 24
[AR2-GigabitEthernet0/0/0]q
[AR2]int g0/0/1
[AR2-GigabitEthernet0/0/1]ip add 192.168.2.2 24
[AR2-GigabitEthernet0/0/1]q
[AR2]ospf 1 router-id 2.2.2.2
[AR2-ospf-1]area 0
[AR2-ospf-1-area-0.0.0.0]network 192.168.1.2 0.0.0.0
[AR2-ospf-1-area-0.0.0.0]network 192.168.2.2 0.0.0.0
```

配置 AR3 的路由。配置如下：

```
[Huawei]sy AR3
[AR3]int g0/0/0
[AR3-GigabitEthernet0/0/0]ip add 192.168.2.3 24
[AR3-GigabitEthernet0/0/0]q
[AR3]ospf 1 router-id 3.3.3.3
[AR3-ospf-1]area 0
[AR3-ospf-1-area-0.0.0.0]network 192.168.2.3 0.0.0.0
[AR3]user-interface vty 0 4
//进入路由器的线路模式，开通虚通道。
[AR3-ui-vty0-4]authentication-mode password
Please configure the login password (maximum length 16):123456
//设置虚通道的密码。
[AR3-ui-vty0-4]user privilege level 15
```

（3）在 AR1 上打开命令行，输入"telnet 192.168.2.3"，出现输入密码的提示后键入预先设置好的密码"123456"。注意该密码在界面上是看不到的，这也是一种安全保护。之后就可以进入路由器的配置界面了，如图 6-3 所示。

图 6-3　测试结果

实验二：动态主机配置协议（DHCP）实验

1. 实验目的

（1）理解 DHCP 的含义。
（2）掌握 DHCP 服务器的配置。

2. 基础知识

动态主机配置协议（Dynamic Host Configuration Protocol，DHCP）通常被应用在局域网环境中，由服务器控制一段 IP 地址范围，对 DHCP 客户主机集中地管理、分配 IP 地址，使客户主机动态地获得 IP 地址、网关、DNS 服务器地址等网络参数。

DHCP 的优点如下。
（1）可以减轻网络管理员的负担。
（2）可以提升 IP 地址的使用率。
（3）可以和其他 IP 地址共存，如静态分配的 IP 地址。

更多详细内容请参考教材《计算机网络（第 8 版）》6.6 节。

3. 实验流程

实验流程如图 6-4 所示。

图 6-4　实验流程图

4. 实验步骤

（1）布置拓扑。如图 6-5 所示，网络中共划分了三个网段，并且设置了一台服务器，其 IP 地址静态划分，其余主机的网络参数自动获取。本实验的 DHCP 服务器为 AR1，但由于

AR1 与主机并不在相同网段,因此需要路由器 AR2 作为中继代为请求 DHCP 服务。其 IP 地址配置如表 6-1 所示。

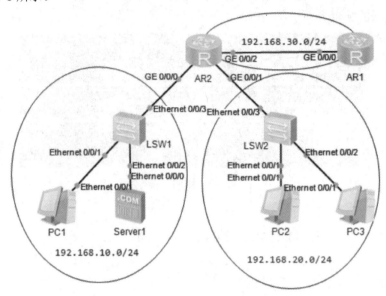

图 6-5 实验拓扑图

表 6-1 各设备的 IP 地址配置

设备名称	接口	IP 地址	网关
AR1	GE0/0/0	192.168.30.2/24	
AR2	GE0/0/0	192.168.10.254/24	
	GE0/0/1	192.168.20.254/24	
	GE0/0/2	192.168.30.1/24	
PC1	Ethernet0/0/1	自动获取	自动获取
Server1	Ethernet0/0/0	192.168.10.1/24	192.168.10.254
PC2	Ethernet0/0/1	自动获取	自动获取
PC3	Ethernet0/0/1	自动获取	自动获取

(2)配置路由。

配置 AR2 的路由。配置如下:

```
[Huawei]sy AR2
[AR2]int g0/0/0
[AR2-GigabitEthernet0/0/0]ip add 192.168.10.254 24
[AR2-GigabitEthernet0/0/0]q
[AR2]int g0/0/1
[AR2-GigabitEthernet0/0/1]ip add 192.168.20.254 24
[AR2-GigabitEthernet0/0/1]q
[AR2]int g0/0/2
```

```
[AR2-GigabitEthernet0/0/2]ip add 192.168.30.1 24
[AR2-GigabitEthernet0/0/2]q
[AR2]ospf 1 router-id 2.2.2.2
[AR2-ospf-1]area 0
[AR2-ospf-1-area-0.0.0.0]network 192.168.10.0 0.0.0.255
[AR2-ospf-1-area-0.0.0.0]network 192.168.20.0 0.0.0.255
[AR2-ospf-1-area-0.0.0.0]network 192.168.30.1 0.0.0.0
```

配置 AR1 的路由。配置如下：

```
[Huawei]sy AR1
[AR1]int g0/0/0
[AR1-GigabitEthernet0/0/0]ip add 192.168.30.2 24
[AR1-GigabitEthernet0/0/0]q
[AR1]ospf 1 router-id 1.1.1.1
[AR1-ospf-1]area 0
[AR1-ospf-1-area-0.0.0.0]network 192.168.30.2 0.0.0.0
```

（3）配置中继代理和 DHCP 服务器。

配置 AR2。配置如下：

```
[AR2]dhcp enable
Info: The operation may take a few seconds. Please wait for a moment.done.
[AR2]int g0/0/0
[AR2-GigabitEthernet0/0/0]dhcp select relay
[AR2-GigabitEthernet0/0/0]dhcp relay server-ip 192.168.30.2
[AR2-GigabitEthernet0/0/0]q
[AR2]int g0/0/1
[AR2-GigabitEthernet0/0/1]dhcp select relay
[AR2-GigabitEthernet0/0/1]dhcp relay server-ip 192.168.30.2
[AR2-GigabitEthernet0/0/1]q
//该接口作为 DHCP 中继，为该网段的主机指定上级 DHCP 服务器的地址。
```

配置 AR1。配置如下：

```
[AR1]dhcp enable
Info: The operation may take a few seconds. Please wait for a moment.done.
//全局开启 DHCP 功能。
[AR1]ip pool VLAN 10
Info: It's successful to create an IP address pool.
[AR1-ip-pool-VLAN 10]gateway-list 192.168.10.254
```

```
[AR1-ip-pool-VLAN 10]network 192.168.10.0 mask 24
//配置 VLAN 10 地址池的地址及网关。
[AR1-ip-pool-VLAN 10]q
[AR1]ip pool VLAN 20
Info: It's successful to create an IP address pool.
[AR1-ip-pool-VLAN 20]gateway-list 192.168.20.254
[AR1-ip-pool-VLAN 20]network 192.168.20.0 mask 24
[AR1-ip-pool-VLAN 20]excluded-ip-address 192.168.20.1 192.168.20.100
//从地址池中排除部分地址，这些地址作为保留，不被自动分配。
[AR1]int g0/0/0
[AR1-GigabitEthernet0/0/0]dhcp select global
[AR1-GigabitEthernet0/0/0]q
```

路由器中共配置了两个地址池，分别用于给 VLAN 10 和 VLAN 20 分配网络参数。

（4）验证主机自动获取 IP 地址。

以 PC1 为例测试，结果如图 6-6 所示，可以看到 PC1 能够自动获得 IP 地址。

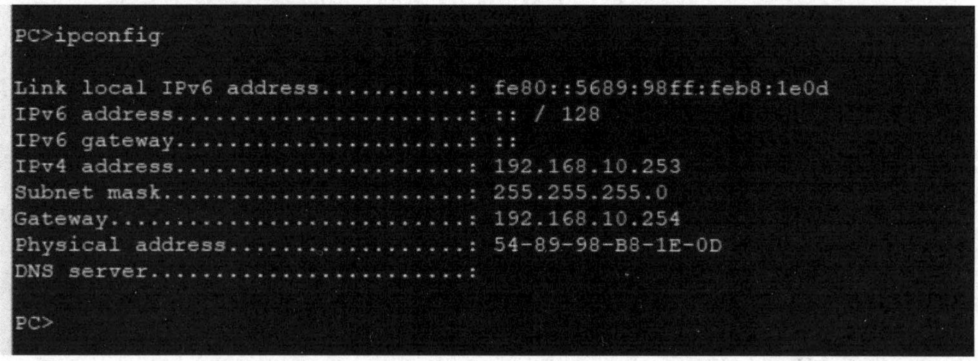

```
PC>ipconfig

Link local IPv6 address...........: fe80::5689:98ff:feb8:1e0d
IPv6 address....................: :: / 128
IPv6 gateway....................: ::
IPv4 address....................: 192.168.10.253
Subnet mask.....................: 255.255.255.0
Gateway.........................: 192.168.10.254
Physical address................: 54-89-98-B8-1E-0D
DNS server......................:

PC>
```

图 6-6　测试结果

读者可以使用 Wireshark 抓取数据包，进一步观察 DHCP 服务的更多细节。

第7章 校园网综合实验

实验一：校园网综合实验1

1. 项目设计

具备一定规模的局域网通常采用核心层、汇聚层、接入层的三层设计模型。

通常将网络中直接面向用户连接或访问网络的部分称为接入层，接入层设备一般可以使用集线器或者二层交换机，具有低成本和高接口密度的特性。汇聚层连接接入层和核心层，多采用三层交换机，是接入层交换机的汇聚点，它处理来自接入层设备的所有通信流量，需要更高的性能和交换速率，一些访问策略也经常被放置在这里。网络主干部分称为核心层，核心层的主要作用是高速转发通信，并提供可靠的骨干传输结构，因此核心层交换机应拥有更高的可靠性、性能和吞吐量。

校园中有两栋教学楼、一栋行政楼、一栋宿舍楼和一栋信息中心楼。信息中心楼是校园网络设备的中心，由此接入互联网。信息中心楼设置了三台核心交换机，服务器直接连接在一台核心交换机上。两栋教学楼、行政楼和宿舍楼中各设置一台二层交换机，并通过三层交换机连接到信息中心楼。三层交换机作为汇聚层交换机，二层交换机作为接入层交换机。

路由协议采用OSPF。为限制LSA的通告范围，提高网络性能，校园网划分了3个区域，核心设备运行在area 0，LSW1连接AR1的链路在area 1，LSW2连接AR1的链路在area 2。

由于受模拟器功能及本书篇幅限制，因此校园网设计在建筑物和部门单位上均进行了简化，核心交换机使用S5700三层交换机，链路连接均使用快速以太网口。本校园网设计仅用于做逻辑上的验证。

2. 实验流程

实验流程如图7-1所示。

图7-1 实验流程图

3. 实验步骤

（1）按图7-2所示布置拓扑，并按表7-1配置各设备的IP地址。

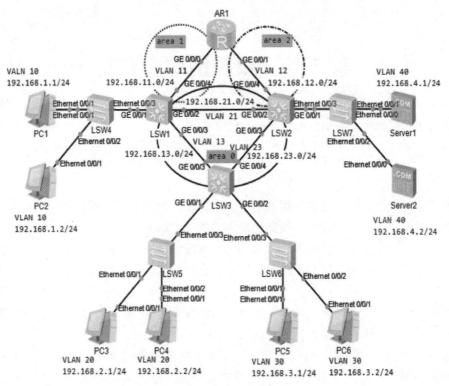

图 7-2 实验拓扑图

表 7-1 各设备的 IP 地址配置

建筑物	设备名称	接口	VLAN	IP 地址	区域（area）
信息中心楼	AR1	GE0/0/0	11	192.168.11.11/24	1
		GE0/0/1	12	192.168.12.1/24	2
	LSW1（三层交换机）	VLAN 10		192.168.1.254/24	0
		VLAN 11		192.168.11.1/24	1
		VLAN 13		192.168.13.1/24	0
		VLAN 21		192.168.21.1/24	0
		GE0/0/1	10		
		GE0/0/2	21		
		GE0/0/3	13		
		GE0/0/4	11		
	LSW2（三层交换机）	VLAN 12		192.168.12.2/24	2
		VLAN 21		192.168.21.2/24	0
		VLAN 23		192.168.23.2/24	0
		VLAN 40		192.168.4.254/24	0
		GE0/0/1	40		
		GE0/0/2	21		
		GE0/0/3	23		
		GE0/0/4	12		

续表

建筑物	设备名称	接口	VLAN	IP 地址	区域（area）
信息中心楼	LSW3（三层交换机）	VLAN 13		192.168.13.3/24	0
		VLAN 20		192.168.2.254/24	0
		VLAN 23		192.168.23.3/24	0
		VLAN 30		192.168.3.254/24	0
		GE0/0/1	20		
		GE0/0/2	30		
		GE0/0/3	13		
		GE0/0/4	23		
宿舍楼	LSW4	VLAN 10			
		Ethernet0/0/1	10		
		Ethernet0/0/2	10		
		Ethernet0/0/3	10		
	PC1			192.168.1.1/24	
	PC2			192.168.1.2/24	
教学楼 1	LSW5	VLAN 20			
		Ethernet0/0/1	20		
		Ethernet0/0/2	20		
		Ethernet0/0/3	20		
	PC3			192.168.2.1/24	
	PC4			192.168.2.2/24	
教学楼 2	LSW6	VLAN 30			
		Ethernet0/0/1	30		
		Ethernet0/0/2	30		
		Ethernet0/0/3	30		
	PC5			192.168.3.1/24	
	PC6			192.168.3.2/24	
行政楼	LSW7	VLAN 40			
		Ethernet0/0/1	40		
		Ethernet0/0/2	40		
		Ethernet0/0/3	40		
	Server1			192.168.4.1/24	
	Server2			192.168.4.2/24	

（2）设备配置。

配置 AR1 的路由。配置如下：

```
[Huawei]sy AR1
[AR1]int g0/0/0
[AR1-GigabitEthernet0/0/0]ip address 192.168.11.11 255.255.255.0
[AR1-GigabitEthernet0/0/0]int g0/0/1
```

```
[AR1-GigabitEthernet0/0/1]ip address 192.168.12.1 255.255.255.0
[AR1-GigabitEthernet0/0/1]q
[AR1]ospf 1 router-id 1.1.1.1
[AR1-ospf-1] area 0.0.0.1
[AR1-ospf-1-area-0.0.0.1]network 192.168.11.11 0.0.0.0
[AR1-ospf-1-area-0.0.0.1]area 0.0.0.2
[AR1-ospf-1-area-0.0.0.2]network 192.168.12.1 0.0.0.0
```

配置核心交换机 LSW1。配置如下:

```
[Huawei]sysname LSW1
[LSW1]vlan batch 10 11 13 21
Info: This operation may take a few seconds. Please wait for a moment...done.
[LSW1]stp disable
[LSW1]interface Vlanif10
[LSW1-Vlanif10]ip address 192.168.1.254 255.255.255.0
[LSW1] interface Vlanif11
[LSW1-Vlanif11]ip address 192.168.11.1 255.255.255.0
[LSW1]interface Vlanif13
[LSW1-Vlanif13]ip address 192.168.13.1 255.255.255.0
[LSW1] interface Vlanif21
[LSW1-Vlanif21]ip address 192.168.21.1 255.255.255.0
[LSW1-Vlanif21]interface GigabitEthernet0/0/1
[LSW1-GigabitEthernet0/0/1]port link-type access
[LSW1-GigabitEthernet0/0/1]port default vlan 10
[LSW1-GigabitEthernet0/0/1]interface GigabitEthernet0/0/2
[LSW1-GigabitEthernet0/0/2]port link-type access
[LSW1-GigabitEthernet0/0/2]port default vlan 21
[LSW1-GigabitEthernet0/0/2]interface GigabitEthernet0/0/3
[LSW1-GigabitEthernet0/0/3]port link-type access
[LSW1-GigabitEthernet0/0/3]port default vlan 13
[LSW1-GigabitEthernet0/0/3]interface GigabitEthernet0/0/4
[LSW1-GigabitEthernet0/0/4]port link-type access
[LSW1-GigabitEthernet0/0/4]port default vlan 11
[LSW1-GigabitEthernet0/0/4]q
[LSW1]ospf 1 router-id 1.1.1.1
[LSW1-ospf-1]area 0.0.0.0
[LSW1-ospf-1-area-0.0.0.0]network 192.168.13.1 0.0.0.0
[LSW1-ospf-1-area-0.0.0.0]network 192.168.21.1 0.0.0.0
```

```
[LSW1-ospf-1-area-0.0.0.0]network 192.168.1.0 0.0.0.255
[LSW1-ospf-1-area-0.0.0.0]area 0.0.0.1
[LSW1-ospf-1-area-0.0.0.1]network 192.168.11.1 0.0.0.0
```

配置核心交换机 LSW2。配置如下:

```
[Huawei]sysname LSW2
[LSW2]vlan batch 12 21 23 40
Info: This operation may take a few seconds. Please wait for a moment...done.
[LSW2]stp disable
[LSW2]interface Vlanif 12
[LSW2-Vlanif12]ip address 192.168.12.2 255.255.255.0
[LSW2-Vlanif12]interface Vlanif21
[LSW2-Vlanif21]ip address 192.168.21.2 255.255.255.0
[LSW2-Vlanif21]interface Vlanif23
[LSW2-Vlanif23]ip address 192.168.23.2 255.255.255.0
[LSW2-Vlanif23]interface Vlanif40
[LSW2-Vlanif40]ip address 192.168.4.254 255.255.255.0
[LSW2-Vlanif40]q
[LSW2]interface g0/0/1
[LSW2-GigabitEthernet0/0/1]port link-type access
[LSW2-GigabitEthernet0/0/1]port default vlan 40
[LSW2-GigabitEthernet0/0/1]interface GigabitEthernet0/0/2
[LSW2-GigabitEthernet0/0/2]port link-type access
[LSW2-GigabitEthernet0/0/2]port default vlan 21
[LSW2-GigabitEthernet0/0/2]interface GigabitEthernet0/0/3
[LSW2-GigabitEthernet0/0/3]port link-type access
[LSW2-GigabitEthernet0/0/3]port default vlan 23
[LSW2-GigabitEthernet0/0/3]interface GigabitEthernet0/0/4
[LSW2-GigabitEthernet0/0/4]port link-type access
[LSW2-GigabitEthernet0/0/4]port default vlan 12
[LSW2-GigabitEthernet0/0/4]q
[LSW2]ospf 1 router-id 2.2.2.2
[LSW2-ospf-1]area 0
[LSW2-ospf-1-area-0.0.0.0]network 192.168.4.0 0.0.0.255
[LSW2-ospf-1-area-0.0.0.0]network 192.168.21.2 0.0.0.0
[LSW2-ospf-1-area-0.0.0.0]network 192.168.23.2 0.0.0.0
[LSW2-ospf-1-area-0.0.0.0]area 2
[LSW2-ospf-1-area-0.0.0.2]network 192.168.12.2 0.0.0.0
```

配置核心交换机LSW3。配置如下：

```
[Huawei]sysname LSW3
[LSW3]vlan batch 13 20 23 30
Info: This operation may take a few seconds. Please wait for a moment...done.
[LSW3]stp disable
[LSW3]interface Vlanif13
[LSW3-Vlanif13]ip address 192.168.13.3 255.255.255.0
[LSW3-Vlanif13]interface Vlanif20
[LSW3-Vlanif20]ip address 192.168.2.254 255.255.255.0
[LSW3-Vlanif20]interface Vlanif23
[LSW3-Vlanif23]ip address 192.168.23.3 255.255.255.0
[LSW3-Vlanif23]interface Vlanif30
[LSW3-Vlanif30]ip address 192.168.3.254 255.255.255.0
[LSW3-Vlanif30]q
[LSW3]interface GigabitEthernet0/0/1
[LSW3-GigabitEthernet0/0/1]port link-type access
[LSW3-GigabitEthernet0/0/1]port default vlan 20
[LSW3-GigabitEthernet0/0/1]interface GigabitEthernet0/0/2
[LSW3-GigabitEthernet0/0/2]port link-type access
[LSW3-GigabitEthernet0/0/2]port default vlan 30
[LSW3-GigabitEthernet0/0/2]interface GigabitEthernet0/0/3
[LSW3-GigabitEthernet0/0/3]port link-type access
[LSW3-GigabitEthernet0/0/3]port default vlan 13
[LSW3-GigabitEthernet0/0/3]interface GigabitEthernet0/0/4
[LSW3-GigabitEthernet0/0/4]port link-type access
[LSW3-GigabitEthernet0/0/4]port default vlan 23
[LSW3-GigabitEthernet0/0/4]q
[LSW3]ospf 1 router-id 3.3.3.3
[LSW3-ospf-1]area 0
[LSW3-ospf-1-area-0.0.0.0]network 192.168.2.0 0.0.0.255
[LSW3-ospf-1-area-0.0.0.0]network 192.168.3.0 0.0.0.255
[LSW3-ospf-1-area-0.0.0.0]network 192.168.13.3 0.0.0.0
[LSW3-ospf-1-area-0.0.0.0]network 192.168.23.3 0.0.0.0
```

配置汇聚交换机LSW4。配置如下：

```
[Huawei]sysname LSW4
[LSW4]vlan 10
Info: This operation may take a few seconds. Please wait for a moment...done.
```

```
[LSW4]interface Ethernet0/0/1
[LSW4-Ethernet0/0/1]port link-type access
[LSW4-Ethernet0/0/1]port default vlan 10
[LSW4-Ethernet0/0/1]interface Ethernet0/0/2
[LSW4-Ethernet0/0/2]port link-type access
[LSW4-Ethernet0/0/2]port default vlan 10
[LSW4-Ethernet0/0/2]interface Ethernet0/0/3
[LSW4-Ethernet0/0/3]port link-type access
[LSW4-Ethernet0/0/3]port default vlan 10
```

配置汇聚交换机LSW5。配置如下:

```
[Huawei]sysname LSW5
[LSW5]vlan batch 20
Info: This operation may take a few seconds. Please wait for a moment...done.
[LSW5]interface Ethernet0/0/1
[LSW5-Ethernet0/0/1]port link-type access
[LSW5-Ethernet0/0/1]port default vlan 20
[LSW5-Ethernet0/0/1]interface Ethernet0/0/2
[LSW5-Ethernet0/0/2]port link-type access
[LSW5-Ethernet0/0/2]port default vlan 20
[LSW5-Ethernet0/0/2]interface Ethernet0/0/3
[LSW5-Ethernet0/0/3]port link-type access
[LSW5-Ethernet0/0/3]port default vlan 20
```

配置汇聚交换机LSW6。配置如下:

```
[Huawei]sysname LSW6
[LSW6]vlan batch 30
Info: This operation may take a few seconds. Please wait for a moment...done.
[LSW6]interface Ethernet0/0/1
[LSW6-Ethernet0/0/1]port link-type access
[LSW6-Ethernet0/0/1]port default vlan 30
[LSW6-Ethernet0/0/1]interface Ethernet0/0/2
[LSW6-Ethernet0/0/2]port link-type access
[LSW6-Ethernet0/0/2]port default vlan 30
[LSW6-Ethernet0/0/2]interface Ethernet0/0/3
[LSW6-Ethernet0/0/3]port link-type access
[LSW6-Ethernet0/0/3]port default vlan 30
```

配置汇聚交换机LSW7。配置如下:

```
[LSW7]sysname LSW7
[LSW7]vlan batch 40
Info: This operation may take a few seconds. Please wait for a moment...done.
[LSW7]interface Ethernet0/0/1
[LSW7-Ethernet0/0/1]port link-type access
[LSW7-Ethernet0/0/1]port default vlan 40
[LSW7-Ethernet0/0/1]interface Ethernet0/0/2
[LSW7-Ethernet0/0/2]port link-type access
[LSW7-Ethernet0/0/2]port default vlan 40
[LSW7-Ethernet0/0/2]interface Ethernet0/0/3
[LSW7-Ethernet0/0/3]port link-type access
[LSW7-Ethernet0/0/3]port default vlan 40
```

其他设备的配置请读者自行练习。

（3）经验证，主机都可相互 ping 通。

（4）查看路由表。

查看核心交换机 LSW1 的路由表，结果如下：

```
<LSW1>display ip routing-table
Route Flags: R - relay, D - download to fib
------------------------------------------------------------------------------
Routing Tables: Public
         Destinations : 15      Routes : 16

Destination/Mask     Proto    Pre  Cost    Flags   NextHop         Interface

       127.0.0.0/8   Direct   0    0       D       127.0.0.1       InLoopBack0
       127.0.0.1/32  Direct   0    0       D       127.0.0.1       InLoopBack0
     192.168.1.0/24  Direct   0    0       D       192.168.1.254   Vlanif10
   192.168.1.254/32  Direct   0    0       D       127.0.0.1       Vlanif10
     192.168.2.0/24  OSPF     10   2       D       192.168.13.3    Vlanif13
     192.168.3.0/24  OSPF     10   2       D       192.168.13.3    Vlanif13
     192.168.4.0/24  OSPF     10   2       D       192.168.21.2    Vlanif21
    192.168.11.0/24  Direct   0    0       D       192.168.11.1    Vlanif11
    192.168.11.1/32  Direct   0    0       D       127.0.0.1       Vlanif11
    192.168.12.0/24  OSPF     10   2       D       192.168.21.2    Vlanif21
    192.168.13.0/24  Direct   0    0       D       192.168.13.1    Vlanif13
    192.168.13.1/32  Direct   0    0       D       127.0.0.1       Vlanif13
    192.168.21.0/24  Direct   0    0       D       192.168.21.1    Vlanif21
```

192.168.21.1/32	Direct	0	0	D	127.0.0.1	Vlanif21
192.168.23.0/24	OSPF	10	2	D	192.168.13.3	Vlanif13
	OSPF	10	2	D	192.168.21.2	Vlanif21

查看核心交换机 LSW2 的路由表，结果如下：

```
<LSW2>display ip routing-table
Route Flags: R - relay, D - download to fib
------------------------------------------------------------------------
Routing Tables: Public
        Destinations : 15        Routes : 16

Destination/Mask      Proto   Pre  Cost     Flags   NextHop         Interface

      127.0.0.0/8     Direct  0    0          D     127.0.0.1       InLoopBack0
      127.0.0.1/32    Direct  0    0          D     127.0.0.1       InLoopBack0
    192.168.1.0/24    OSPF    10   2          D     192.168.21.1    Vlanif21
    192.168.2.0/24    OSPF    10   2          D     192.168.23.3    Vlanif23
    192.168.3.0/24    OSPF    10   2          D     192.168.23.3    Vlanif23
    192.168.4.0/24    Direct  0    0          D     192.168.4.254   Vlanif40
  192.168.4.254/32    Direct  0    0          D     127.0.0.1       Vlanif40
   192.168.11.0/24    OSPF    10   2          D     192.168.21.1    Vlanif21
   192.168.12.0/24    Direct  0    0          D     192.168.12.2    Vlanif12
   192.168.12.2/32    Direct  0    0          D     127.0.0.1       Vlanif12
   192.168.13.0/24    OSPF    10   2          D     192.168.23.3    Vlanif23
                      OSPF    10   2          D     192.168.21.1    Vlanif21
   192.168.21.0/24    Direct  0    0          D     192.168.21.2    Vlanif21
   192.168.21.2/32    Direct  0    0          D     127.0.0.1       Vlanif21
   192.168.23.0/24    Direct  0    0          D     192.168.23.2    Vlanif23
   192.168.23.2/32    Direct  0    0          D     127.0.0.1       Vlanif23
```

实验二：校园网综合实验 2

1. 项目设计

校园的左半部分为主校区，右半部分为分校区，主校区（area 0）的路由协议采用 OSPF，分校区的路由协议采用 RIP，主校区与分校区之间通过路由重分发使双方互通。主校区有教学楼、宿舍楼、行政楼、图书馆各一栋，分校区有一栋教学楼。主校区的二层交换机与三层交换机通过 STP 防环，并配置 LSW2 为 VLAN 10 的根桥，LSW3 为 VLAN 20 的根桥。LSW2

与LSW3之间使用链路聚合技术，且配置为手工负载分担模式。LSW2与LSW3配置了VRRP来增强拓扑的冗余性，使教学楼与宿舍楼向外访问时流量优先走LSW2，行政楼与图书馆向外访问时流量优先走LSW3。

由于受模拟器功能及本书篇幅限制，因此校园网设计在建筑物和部门单位上均进行了简化，核心交换机使用S5700三层交换机，链路连接均使用快速以太网口。本校园网设计仅用于做逻辑上的验证。

2. 实验流程

实验流程如图7-3所示。

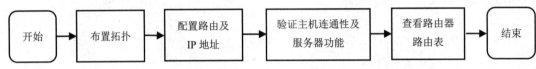

图7-3 实验流程图

3. 实验步骤

（1）按图7-4所示布置拓扑，并按表7-2配置各设备的IP地址。

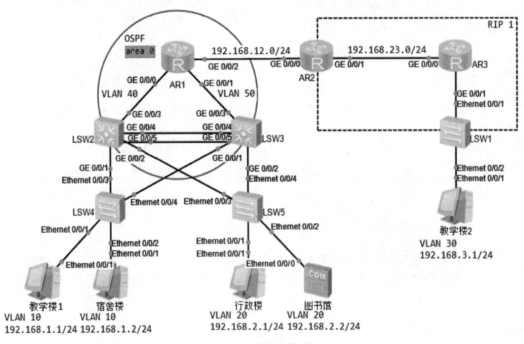

图7-4 实验拓扑图

表7-2 各设备的IP地址配置

校区	设备名称	接口	VLAN	IP地址	区域（area）	RIP
主校区	AR1	GE0/0/0	40	192.168.21.1/24	0	
		GE0/0/1	50	192.168.31.1/24	0	
		GE0/0/2		192.168.12.1/24	0	

续表

校区	设备名称	接口	VLAN	IP 地址	区域（area）	RIP
主校区	AR2	GE0/0/0		192.168.12.2/24	0	
		GE0/0/1		192.168.23.2/24		1
	LSW2	GE0/0/1	10、20、40			
		GE0/0/2	10、20、40			
		GE0/0/3	40			
		Eth-Trunk 1	10、20、40			
		VLAN 10		192.168.1.253/24	0	
			虚拟 IP 地址：192.168.1.254/24			
		VLAN 20		192.168.2.253/24	0	
			虚拟 IP 地址：192.168.2.254/24			
		VLAN 40		192.168.21.2/24	0	
	LSW3	GE0/0/1	10、20、50			
		GE0/0/2	10、20、50			
		GE0/0/3	50			
		Eth-Trunk 1	10、20、50			
		VLAN 10		192.168.1.252/24		
			虚拟 IP 地址：192.168.1.254/24			
		VLAN 20		192.168.2.252/24		
			虚拟 IP 地址：192.168.2.254/24			
		VLAN 50		192.168.31.3/24		
	LSW4	Ethernet0/0/1	10			
		Ethernet0/0/2	10			
		Ethernet0/0/3	10			
		Ethernet0/0/4	10			
	LSW5	Ethernet0/0/1	20			
		Ethernet0/0/2	20			
		Ethernet0/0/3	20			
		Ethernet0/0/4	20			
分校区	AR3	GE0/0/0		192.168.23.3/24		1
		GE0/0/1		192.168.3.254/24		1
	LSW1	Ethernet0/0/1	30			
		Ethernet0/0/2	30			

（2）设备配置。

配置路由器 AR1。配置如下：

```
[AR1]sysname AR1
[AR1]interface GigabitEthernet0/0/0
[AR1-GigabitEthernet0/0/0]ip address 192.168.21.1 255.255.255.0
[AR1-GigabitEthernet0/0/0]interface GigabitEthernet0/0/1
```

```
[AR1-GigabitEthernet0/0/1]ip address 192.168.31.1 255.255.255.0
[AR1-GigabitEthernet0/0/1]interface GigabitEthernet0/0/2
[AR1-GigabitEthernet0/0/2]ip address 192.168.12.1 255.255.255.0
[AR1-GigabitEthernet0/0/2]q
[AR1]ospf 1 router-id 1.1.1.1
[AR1-ospf-1]area 0.0.0.0
[AR1-ospf-1-area-0.0.0.0]network 192.168.12.1 0.0.0.0
[AR1-ospf-1-area-0.0.0.0]network 192.168.21.1 0.0.0.0
[AR1-ospf-1-area-0.0.0.0]network 192.168.31.1 0.0.0.0
```

配置路由器 AR2。配置如下：

```
[AR2]sysname AR2
[AR2]interface GigabitEthernet0/0/0
[AR2-GigabitEthernet0/0/0]ip address 192.168.12.2 255.255.255.0
[AR2-GigabitEthernet0/0/0]interface GigabitEthernet0/0/1
[AR2-GigabitEthernet0/0/1]ip address 192.168.23.2 255.255.255.0
[AR2-GigabitEthernet0/0/1]q
[AR2]ospf 1 router-id 22.22.22.22
[AR2-ospf-1]import-route rip 1
[AR2-ospf-1]area 0.0.0.0
[AR2-ospf-1-area-0.0.0.0]network 192.168.12.2 0.0.0.0
[AR2-ospf-1-area-0.0.0.0]q
[AR2]rip 1
[AR2-rip-1]network 192.168.23.0
[AR2-rip-1]import-route ospf 1
```

配置路由器 AR3。配置如下：

```
[AR3]sysname AR3
[AR3]interface GigabitEthernet0/0/0
[AR3-GigabitEthernet0/0/0]ip address 192.168.23.3 255.255.255.0
[AR3-GigabitEthernet0/0/0]interface GigabitEthernet0/0/1
[AR3-GigabitEthernet0/0/1]ip address 192.168.3.254 255.255.255.0
[AR3-GigabitEthernet0/0/1]q
[AR3]rip 1
[AR3-rip-1]network 192.168.23.0
[AR3-rip-1]network 192.168.3.0
```

配置交换机 LSW1。配置如下：

```
[LSW1]sysname LSW1
[LSW1]vlan batch 30
[LSW1]interface Ethernet0/0/1
[LSW1-Ethernet0/0/1]port link-type access
[LSW1-Ethernet0/0/1]port default vlan 30
[LSW1-Ethernet0/0/1]interface Ethernet0/0/2
[LSW1-Ethernet0/0/2]port link-type access
[LSW1-Ethernet0/0/2]port default vlan 30
```

配置交换机 LSW2。配置如下：

```
[LSW2]sysname LSW2
[LSW2]vlan batch 10 20 40
[LSW2]stp region-configuration
[LSW2-mst-region]region-name jiaoxue
[LSW2-mst-region]instance 10 vlan 10
[LSW2-mst-region]instance 20 vlan 20
[LSW2-mst-region]active region-configuration
[LSW2-mst-region]q
[LSW2]stp instance 10 root primary
[LSW2]stp instance 20 root secondary
[LSW2]interface Vlanif10
[LSW2-Vlanif10]ip address 192.168.1.253 255.255.255.0
[LSW2-Vlanif10]vrrp vrid 1 virtual-ip 192.168.1.254      //配置VRRP虚拟IP地址。
[LSW2-Vlanif10]vrrp vrid 1 priority 120                  //配置VRRP优先级。
[LSW2-Vlanif10]interface Vlanif20
[LSW2-Vlanif20]ip address 192.168.2.253 255.255.255.0
[LSW2-Vlanif20]vrrp vrid 2 virtual-ip 192.168.2.254      //配置VRRP虚拟IP地址。
[LSW2-Vlanif20]interface Vlanif40
[LSW2-Vlanif40]ip address 192.168.21.2 255.255.255.0
[LSW2-Vlanif40]interface Eth-Trunk1
[LSW2-Eth-Trunk1]port link-type trunk
[LSW2-Eth-Trunk1]undo port trunk allow-pass vlan 1
[LSW2-Eth-Trunk1]port trunk allow-pass vlan 2 to 4094
[LSW2-Eth-Trunk1]interface GigabitEthernet0/0/1
[LSW2-GigabitEthernet0/0/1]port link-type trunk
[LSW2-GigabitEthernet0/0/1]undo port trunk allow-pass vlan 1
[LSW2-GigabitEthernet0/0/1]port trunk allow-pass vlan 2 to 4094
[LSW2-GigabitEthernet0/0/1]interface GigabitEthernet0/0/2
```

```
[LSW2-GigabitEthernet0/0/2]port link-type trunk
[LSW2-GigabitEthernet0/0/2]undo port trunk allow-pass vlan 1
[LSW2-GigabitEthernet0/0/2]port trunk allow-pass vlan 2 to 4094
[LSW2-GigabitEthernet0/0/2]interface GigabitEthernet0/0/3
[LSW2-GigabitEthernet0/0/3]port link-type access
[LSW2-GigabitEthernet0/0/3]port default vlan 40
[LSW2-GigabitEthernet0/0/3]interface GigabitEthernet0/0/4
[LSW2-GigabitEthernet0/0/4]eth-trunk 1
[LSW2-GigabitEthernet0/0/4]interface GigabitEthernet0/0/5
[LSW2-GigabitEthernet0/0/5]eth-trunk 1
[LSW2-GigabitEthernet0/0/5]q
[LSW2]ospf 1 router-id 2.2.2.2
[LSW2-ospf-1]area 0.0.0.0
[LSW2-ospf-1-area-0.0.0.0]network 192.168.1.0 0.0.0.255
[LSW2-ospf-1-area-0.0.0.0]network 192.168.2.0 0.0.0.255
[LSW2-ospf-1-area-0.0.0.0]network 192.168.21.2 0.0.0.0
```

配置交换机 LSW3。配置如下：

```
[LSW3]sysname LSW3
[LSW3]vlan batch 10 20 50
[LSW3]stp region-configuration
[LSW3-mst-region]region-name jiaoxue
[LSW3-mst-region]instance 10 vlan 10
[LSW3-mst-region]instance 20 vlan 20
[LSW3-mst-region]active region-configuration
[LSW3-mst-region]q
[LSW3]stp instance 10 root secondary
[LSW3]stp instance 20 root primary
[LSW3]interface Vlanif10
[LSW3-Vlanif10]ip address 192.168.1.252 255.255.255.0
[LSW3-Vlanif10]vrrp vrid 1 virtual-ip 192.168.1.254      //配置VRRP虚拟IP地址。
[LSW3-Vlanif10]interface Vlanif20
[LSW3-Vlanif20]ip address 192.168.2.252 255.255.255.0
[LSW3-Vlanif20]vrrp vrid 2 virtual-ip 192.168.2.254      //配置VRRP虚拟IP地址。
[LSW3-Vlanif20]vrrp vrid 2 priority 120                  //配置VRRP优先级。
[LSW3-Vlanif20]interface Vlanif50
[LSW3-Vlanif50]ip address 192.168.31.3 255.255.255.0
[LSW3-Vlanif50]q
```

```
[LSW3]interface Eth-Trunk1
[LSW3-Eth-Trunk1]port link-type trunk
[LSW3-Eth-Trunk1]undo port trunk allow-pass vlan 1
[LSW3-Eth-Trunk1]port trunk allow-pass vlan 2 to 4094
[LSW3-Eth-Trunk1]interface GigabitEthernet0/0/1
[LSW3-GigabitEthernet0/0/1]port link-type trunk
[LSW3-GigabitEthernet0/0/1]undo port trunk allow-pass vlan 1
[LSW3-GigabitEthernet0/0/1]port trunk allow-pass vlan 2 to 4094
[LSW3-GigabitEthernet0/0/1]interface GigabitEthernet0/0/2
[LSW3-GigabitEthernet0/0/2]port link-type trunk
[LSW3-GigabitEthernet0/0/2]undo port trunk allow-pass vlan 1
[LSW3-GigabitEthernet0/0/2]port trunk allow-pass vlan 2 to 4094
[LSW3-GigabitEthernet0/0/2]interface GigabitEthernet0/0/3
[LSW3-GigabitEthernet0/0/3]port link-type access
[LSW3-GigabitEthernet0/0/3]port default vlan 50
[LSW3-GigabitEthernet0/0/3]interface GigabitEthernet0/0/4
[LSW3-GigabitEthernet0/0/4]eth-trunk 1
[LSW3-GigabitEthernet0/0/4]interface GigabitEthernet0/0/5
[LSW3-GigabitEthernet0/0/5]eth-trunk 1
[LSW3-GigabitEthernet0/0/5]q
[LSW3]ospf 1 router-id 3.3.3.3
[LSW3-ospf-1]area 0.0.0.0
[LSW3-ospf-1-area-0.0.0.0]network 192.168.1.0 0.0.0.255
[LSW3-ospf-1-area-0.0.0.0]network 192.168.2.0 0.0.0.255
[LSW3-ospf-1-area-0.0.0.0]network 192.168.31.3 0.0.0.0
```

配置交换机LSW4。配置如下：

```
[LSW4]sysname LSW4
[LSW4]vlan batch 10
[LSW4]interface Ethernet0/0/1
[LSW4-Ethernet0/0/1]port link-type access
[LSW4-Ethernet0/0/1]port default vlan 10
[LSW4-Ethernet0/0/1]interface Ethernet0/0/2
[LSW4-Ethernet0/0/2]port link-type access
[LSW4-Ethernet0/0/2]port default vlan 10
[LSW4-Ethernet0/0/2]interface Ethernet0/0/3
[LSW4-Ethernet0/0/3]port link-type trunk
[LSW4-Ethernet0/0/3]undo port trunk allow-pass vlan 1
```

```
[LSW4-Ethernet0/0/3]port trunk allow-pass vlan 2 to 4094
[LSW4-Ethernet0/0/3]interface Ethernet0/0/4
[LSW4-Ethernet0/0/4]port link-type trunk
[LSW4-Ethernet0/0/4]undo port trunk allow-pass vlan 1
[LSW4-Ethernet0/0/4]port trunk allow-pass vlan 2 to 4094
```

配置交换机 LSW5。配置如下：

```
[LSW5]sysname LSW5
[LSW5]vlan batch 20
[LSW5]interface Ethernet0/0/1
[LSW5-Ethernet0/0/1]port link-type access
[LSW5-Ethernet0/0/1]port default vlan 20
[LSW5-Ethernet0/0/1]interface Ethernet0/0/2
[LSW5-Ethernet0/0/2]port link-type access
[LSW5-Ethernet0/0/2]port default vlan 20
[LSW5-Ethernet0/0/2]interface Ethernet0/0/3
[LSW5-Ethernet0/0/3]port link-type trunk
[LSW5-Ethernet0/0/3]undo port trunk allow-pass vlan 1
[LSW5-Ethernet0/0/3]port trunk allow-pass vlan 2 to 4094
[LSW5-Ethernet0/0/3]interface Ethernet0/0/4
[LSW5-Ethernet0/0/4]port link-type trunk
[LSW5-Ethernet0/0/4]undo port trunk allow-pass vlan 1
[LSW5-Ethernet0/0/4]port trunk allow-pass vlan 2 to 4094
```

其他设备的配置请读者自行练习。

（3）经验证，主机都可相互 ping 通。

（4）查看路由表。

查看路由器 AR1 的路由表，结果如下：

```
[AR1]display ip routing-table
Route Flags: R - relay, D - download to fib
------------------------------------------------------------------------
Routing Tables: Public
         Destinations : 19        Routes : 21
Destination/Mask      Proto    Pre  Cost  Flags  NextHop         Interface
     127.0.0.0/8      Direct   0    0     D      127.0.0.1       InLoopBack0
     127.0.0.1/32     Direct   0    0     D      127.0.0.1       InLoopBack0
127.255.255.255/32    Direct   0    0     D      127.0.0.1       InLoopBack0
     192.168.1.0/24   OSPF     10   2     D      192.168.21.2    GigabitEthernet0/0/0
```

		OSPF	10	2	D	192.168.31.3	GigabitEthernet0/0/1
192.168.1.254/32		OSPF	10	2	D	192.168.21.2	GigabitEthernet0/0/0
192.168.2.0/24		OSPF	10	2	D	192.168.21.2	GigabitEthernet0/0/0
		OSPF	10	2	D	192.168.31.3	GigabitEthernet0/0/1
192.168.2.254/32		OSPF	10	2	D	192.168.31.3	GigabitEthernet0/0/1
192.168.3.0/24		O_ASE	150	1	D	192.168.12.2	GigabitEthernet0/0/2
192.168.12.0/24		Direct	0	0	D	192.168.12.1	GigabitEthernet0/0/2
192.168.12.1/32		Direct	0	0	D	127.0.0.1	GigabitEthernet0/0/2
192.168.12.255/32		Direct	0	0	D	127.0.0.1	GigabitEthernet0/0/2
192.168.21.0/24		Direct	0	0	D	192.168.21.1	GigabitEthernet0/0/0
192.168.21.1/32		Direct	0	0	D	127.0.0.1	GigabitEthernet0/0/0
192.168.21.255/32		Direct	0	0	D	127.0.0.1	GigabitEthernet0/0/0
192.168.23.0/24		O_ASE	150	1	D	192.168.12.2	GigabitEthernet0/0/2
192.168.31.0/24		Direct	0	0	D	192.168.31.1	GigabitEthernet0/0/1
192.168.31.1/32		Direct	0	0	D	127.0.0.1	GigabitEthernet0/0/1
192.168.31.255/32		Direct	0	0	D	127.0.0.1	GigabitEthernet0/0/1
255.255.255.255/32		Direct	0	0	D	127.0.0.1	InLoopBack0

查看路由器 AR2 的路由表，结果如下：

```
[AR2]display ip routing-table
Route Flags: R - relay, D - download to fib
------------------------------------------------------------------------
Routing Tables: Public
         Destinations : 17       Routes : 17
Destination/Mask    Proto   Pre  Cost Flags NextHop         Interface
      127.0.0.0/8   Direct  0    0     D    127.0.0.1       InLoopBack0
      127.0.0.1/32  Direct  0    0     D    127.0.0.1       InLoopBack0
127.255.255.255/32  Direct  0    0     D    127.0.0.1       InLoopBack0
    192.168.1.0/24  OSPF    10   3     D    192.168.12.1    GigabitEthernet0/0/0
  192.168.1.254/32  OSPF    10   3     D    192.168.12.1    GigabitEthernet0/0/0
    192.168.2.0/24  OSPF    10   3     D    192.168.12.1    GigabitEthernet0/0/0
  192.168.2.254/32  OSPF    10   3     D    192.168.12.1    GigabitEthernet0/0/0
    192.168.3.0/24  RIP     100  1     D    192.168.23.3    GigabitEthernet0/0/1
   192.168.12.0/24  Direct  0    0     D    192.168.12.2    GigabitEthernet0/0/0
   192.168.12.2/32  Direct  0    0     D    127.0.0.1       GigabitEthernet0/0/0
 192.168.12.255/32  Direct  0    0     D    127.0.0.1       GigabitEthernet0/0/0
   192.168.21.0/24  OSPF    10   2     D    192.168.12.1    GigabitEthernet0/0/0
   192.168.23.0/24  Direct  0    0     D    192.168.23.2    GigabitEthernet0/0/1
```

```
     192.168.23.2/32    Direct  0    0       D   127.0.0.1      GigabitEthernet0/0/1
   192.168.23.255/32    Direct  0    0       D   127.0.0.1      GigabitEthernet0/0/1
     192.168.31.0/24    OSPF    10   2       D   192.168.12.1   GigabitEthernet0/0/0
  255.255.255.255/32    Direct  0    0       D   127.0.0.1      InLoopBack0
```

查看路由器 AR3 的路由表,结果如下:

```
[AR3]display ip routing-table
Route Flags: R - relay, D - download to fib
------------------------------------------------------------------------------
Routing Tables: Public
         Destinations : 15       Routes : 15
Destination/Mask     Proto   Pre  Cost    Flags NextHop        Interface
      127.0.0.0/8    Direct  0    0       D   127.0.0.1      InLoopBack0
      127.0.0.1/32   Direct  0    0       D   127.0.0.1      InLoopBack0
  127.255.255.255/32 Direct  0    0       D   127.0.0.1      InLoopBack0
     192.168.1.0/24  RIP     100  1       D   192.168.23.2   GigabitEthernet0/0/0
     192.168.2.0/24  RIP     100  1       D   192.168.23.2   GigabitEthernet0/0/0
     192.168.3.0/24  Direct  0    0       D   192.168.3.254  GigabitEthernet0/0/1
   192.168.3.254/32  Direct  0    0       D   127.0.0.1      GigabitEthernet0/0/1
   192.168.3.255/32  Direct  0    0       D   127.0.0.1      GigabitEthernet0/0/1
    192.168.12.0/24  RIP     100  1       D   192.168.23.2   GigabitEthernet0/0/0
    192.168.21.0/24  RIP     100  1       D   192.168.23.2   GigabitEthernet0/0/0
    192.168.23.0/24  Direct  0    0       D   192.168.23.3   GigabitEthernet0/0/0
    192.168.23.3/32  Direct  0    0       D   127.0.0.1      GigabitEthernet0/0/0
  192.168.23.255/32  Direct  0    0       D   127.0.0.1      GigabitEthernet0/0/0
    192.168.31.0/24  RIP     100  1       D   192.168.23.2   GigabitEthernet0/0/0
  255.255.255.255/32 Direct  0    0       D   127.0.0.1      InLoopBack0
```

附录 A eNSP 常见问题及解决办法

问题 1. 关于提示"没有找到 packet.dll, 因此这个应用程序未能启动。"的问题。
由于软件依赖 WinPcap, 因此需要安装 WinPcap 软件。
问题 2. 使用计算机自带的终端查看命令解释时对齐不准确。
单击"menu→tools→options"选项, 在弹出的 Font Config 页面中, 通过 CLI Font 设置一个等宽字体, 如 Consolas。
问题 3. 桥接本地网络最多只能显示 5 个本地网卡, 且本地网卡无法显示名称。
目前尚不支持分列显示多个网卡信息, 最大化窗口后, 可查看更多的网卡信息。
问题 4. 启动路由器和交换机时有时会比较慢, 交换机启动后会出现很多 alarm。
设备启动需要初始化过程, 故需要一定时间, 当启动完成后, 会有些提示信息, 属于正常情况。
问题 5. 设备显示 CPU 占用率高, 是否会影响搭建大型拓扑?
以任务管理器中真实的 CPU 占用率为准, 实际占用率很低。
问题 6. 可不可以做分布式实验?
可以, 请参照帮助文档中的"灵活部署→配置服务端和客户端"章节。
问题 7. 是否支持 Eth-Trunk?
目前只支持 Eth-Trunk, 具体配置可参照安装目录下 Help 文件夹中的产品手册。
问题 8. S3700 交换机默认生成的 GE 接口在视图上显示的是 GE0/0/23, 但在设备 CLI 配置上进入的接口是 GE0/0/1, 视图显示接口错误。
不影响正常的接口使用, 仅是显示问题, 后续会进行修正。
问题 9. 右击设备图标后单击"start"按钮, 进度条会运行到 100%, 但是双击设备, 设备显示长度未变化的一串#号。
设备启动需要初始化过程, 请耐心等待。
问题 10. 当 eNSP 启动时, 会被某些杀毒软件识别为病毒。
属于误报, 可以尝试加入白名单。
问题 11. 报接口绑定错误。
单击"菜单→工具→选项→服务器页面"选项, 将里面的三个接口号分别改成 54012、54013、54014, 然后单击"应用→确定"按钮, 重启 eNSP。如果重启后还是报接口绑定错误, 那么就将这三个数字接着往后加 1, 再试, 一直到不报错为止。报接口绑定错误的原因是 eNSP 需要的接口号被系统占用了。
问题 12. 如何降低 CPU 占用率, 开启 VBox 虚拟化?
查看 eNSP 帮助文档 FAQ——如何解决使用 eNSP 工具时 CPU 利用率过高的问题。
问题 13. 当启动 AR 时, 提示"can not find AR_Base"错误信息。
单击"菜单→工具→注册设备"选项, 将 AR_Base 重新注册。

如果注册的过程中提示无法注册,那么按照下面的步骤进行。

(1)打开 virtualbox,查看是否有 AR_Clone_或 AR_Base 开头的链接,若有则删除。

(2)单击 virtualbox 的"管理→虚拟介质管理"选项,将 AR_Base.vdi 删除,如果提示无法删除,那么先删除 AR_Base.vdi 下面的子链接,再删除 AR_Base.vdi。

(3)打开 eNSP,不要添加任何模拟设备,直接单击"菜单→工具→注册设备"选项,然后进行注册。

(4)如果 AC、AP 设备也提示无法找到,那么也按照上述方法操作,只是删除的链接的名称分别是 WLAN_AC_Base、WLAN_AP_Base。

问题 14. 启动交换机后,命令行在长时间等待后一直在输出"####"。

(1)单击"开始→附件"选项,然后右击"命令提示符"选项,单击"以管理员方式运行"。

(2)在"管理员:命令提示符"窗口中键入"lodctr /R"(Windows 系统)或"lodctr /R:PerfStringBackup.ini"(XP 系统)来重启计数器,然后重启 eNSP。

问题 15. 在启动 AR 的过程中,出现错误码 40,或者启动后,AR 的命令行在长时间等待后一直在输出"####"。

一般可以做如下几项检查。

(1)打开"控制面板→网络和 Internet→网络连接",检查是否存在一个命名为"VirtualBox Host-Only Network"的虚拟网卡,若该虚拟网卡被禁用,则启用。

(2)打开 VBox,选择"管理→全局设定"选项。

反侵权盗版声明

电子工业出版社依法对本作品享有专有出版权。任何未经权利人书面许可，复制、销售或通过信息网络传播本作品的行为；歪曲、篡改、剽窃本作品的行为，均违反《中华人民共和国著作权法》，其行为人应承担相应的民事责任和行政责任，构成犯罪的，将被依法追究刑事责任。

为了维护市场秩序，保护权利人的合法权益，我社将依法查处和打击侵权盗版的单位和个人。欢迎社会各界人士积极举报侵权盗版行为，本社将奖励举报有功人员，并保证举报人的信息不被泄露。

举报电话：（010）88254396；（010）88258888
传　　真：（010）88254397
E-mail：dbqq@phei.com.cn
通信地址：北京市万寿路173信箱
　　　　　电子工业出版社总编办公室
邮　　编：100036